THE COMMONWEA
Joint Chairmen of the
SIR ROBERT ROBINSON,
DEAN ATHELSTAN SPILHAUS, MINNESOTA

ELECTRICAL ENGINEERING
General Editor: N. HILLER

CA
FU

VOI

VOL. I: Mechanics and Heat

VOL. II: Electricity and Magnetism

VOL. III: Waves and Electronics

CALCULATIONS IN FUNDAMENTAL PHYSICS

In Three Volumes

VOLUME II: ELECTRICITY AND MAGNETISM

by

T. HEDDLE, B.SC. (HONS.), A.INST.P.

Senior Lecturer, South East London Technical College

PERGAMON PRESS

Oxford · New York · Toronto

Sydney · Braunschweig

PERGAMON PRESS LTD.,
Headington Hill Hall, Oxford
PERGAMON PRESS INC.,
Maxwell House, Fairview Park, Elmsford, New York 10523
PERGAMON OF CANADA LTD.
207 Queen's Quay West, Toronto 1
PERGAMON PRESS (AUST.) PTY. LTD.,
19a Boundary Street, Rushcutters Bay, N.S.W. 2011, Australia
VIEWEG & SOHN GMBH,
Burgplatz 1, Braunschweig

First edition 1971
Library of Congress Catalog Card No. 72-121012

Printed in Hungary

08 015831 5

CONTENTS

INTRODUCTION

THIS volume is the second of three which form neither a complete textbook of physics nor a mere collection of exercises. It contains a sequence of worked examples and unworked exercises arranged to present a logical development from the basic definitions of physical quantities to the more complex relationships between them.

Each section starts with a worked example in which first a numerical problem is stated, then a general quantitative development of the relevant concepts and equations is usually given, and, finally, the numerical solution is shown in detail. The exercises which follow in each section refer to the particular topic of the worked example or examples which precede them, and do not range widely from this topic.

The data of most questions are stated to only two significant figures, but the data of the reference section are stated to greater accuracy in most cases. Most answers are therefore given to an accuracy of two significant figures, even when a third figure may appear to terminate the number exactly.

The divisions of the subject matter follow fairly traditional lines to help in finding a problem or explanation of a particular point. While such divisions make the portions readily digestible, this is not sufficient in a complete course in physics, which should obviously include considerable amounts of laboratory work, both demonstrations and experiments, and practice in the description and explanation of apparatus and other examples of the embodiment of physical principles. The attempt has been made not to present topics as they might appear in specialist studies of specific branches of the subject, but to show the fundamental physical principles on which applied physics and engineering are founded.

The most elementary work of the book may be described as revision of the physics of pre-O.N.C. courses in science or engineering or of O-level G.C.E. work in physics, while the final level is somewhat above the physics of most O.N.C. and O.N.D. courses in science and engineering. Since syllabuses vary between different colleges, the range of topics is wider and some treatment is deeper than may be necessary for any particular course. The introductory notes should be useful for reference and for further work in H.N.C. and H.N.D. courses.

I gladly acknowledge help from many sources during the years in which this book has grown. Much assistance has been gained from the publications of numerous authors who may recognize ideas that they have expressed originally. Many colleagues have given helpful advice on specific points even unwittingly before the book was started. I regret any errors and obscurities which readers may find, and would welcome their being brought to my attention.

TREVOR HEDDLE

REFERENCE DATA

Property	Unit	Substance			
		Aluminium	Copper	Silver	Water
Density*	$kg\ m^{-3}$	2.7×10^3	8.9×10^3	1.05×10^4	1.00×10^3
Specific heat*	$J\ kg^{-1}\ deg\ C^{-1}$	910	390		4.19×10^3
Electrochemical equivalent	$kg\ C^{-1}$		3.3×10^{-7}	1.118×10^{-6}	
Resistivity (at 20°C)	$\Omega\ m$	2.8×10^{-8}	1.7×10^{-8}		
Temperature coefficient of resistance (at 0°C)	$deg\ C^{-1}$	3.8×10^{-3}	4.3×10^{-3}		

* Temperature variation to be neglected.

General Constants

Standard acceleration of gravity: $g = 9.81\ m\ s^{-2}$.
Charge of one electron: $e = 1.60\times10^{-19}$ C.
Faraday's constant of electrolysis $= 9.65\times10^7$ C per kilogram equivalent.
Avogadro's number $= 6.02\times10^{26}$ molecules per kilomole.
Permeability of free space $\mu_0 = 4\pi\times10^{-7}\ H\ m^{-1}$.
Permittivity of free space $\varepsilon_0 = 8.85\times10^{-12}\ F\ m^{-1}$.

s.t.p. = standard temperature and pressure.
= 0°C and 760 mmHg.
= 273 K and $1.013\times10^5\ N\ m^{-2}$.
1 litre $= 10^{-3}\ m^3$.

CHAPTER 1

CURRENT AND POTENTIAL DIFFERENCE

Flow of Charge

1.1. Worked Example

Two lamps are connected in parallel across a battery of cells. A current of 1.6 A flows through the battery when the current through one lamp is 7.5×10^{18} electrons per second. Calculate the current through the second lamp (a) in amperes, and (b) in electrons per second.

Introduction

Two types of charged particles are involved in the structure of any material: the positively charged nuclei of the atoms and the negatively charged electrons moving around the nuclei. When no current flows in a stationary conductor, the continual orbiting and jostling of these particles leads to no resultant large scale movement of either positive or negative charge one way or another. If, however, an electric cell is connected in a closed circuit of conductors, the cell tends to drive negative particles one way around the circuit and positive the other. The direction of flow of the positive charge is said to be the direction of the current. Yet in most solid conductors, electric currents consist only of those electrons which are free to drift through the solid, while the positive nuclei, with orbiting electrons, remain localized in the solid structure.

Many effects of currents are the same whether due to positive particles drifting in one direction or to negative in the other. Also the magnitudes of currents are more easily measured by their effects than by attempting to count the numbers or rates of flow of the particles. Thus the unit of current, the ampere (A), is determined by a convenient effect, namely the force exerted between parallel wires carrying currents (as in Fig. 1.1). The ampere is that

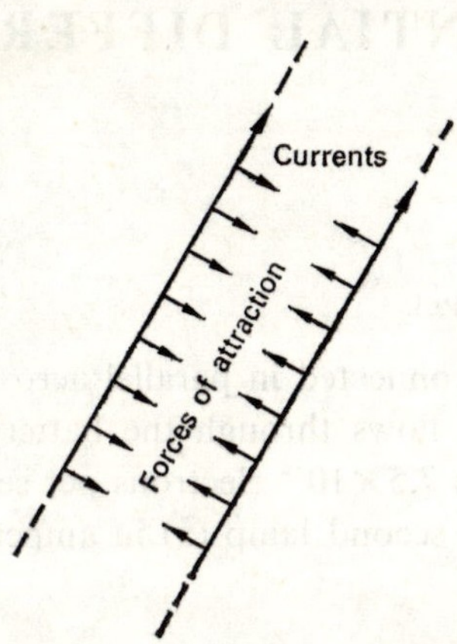

FIG. 1.1. Forces of attraction between current-carrying conductors.

constant current which, in each of two parallel, infinitely long, and straight wires of negligible circular cross-section, spaced one metre apart in a vacuum, produces on each wire a force of 2×10^{-7} newton per metre length. The number 2×10^{-7} makes this definition agree with another referring to curved conductors, but now obsolete.

The unit of charge corresponding to the ampere of current is the coulomb. Thus one coulomb (C) is the resultant charge passed in one second through any full cross-section of a constant current of one ampere. In terms of this unit the very small negative charge of each electron is found to be 1.6×10^{-19} C. Many charges are measured in microcoulombs (μC); 1 μC = 10^{-6} C. In general, the charge Q which passes when a constant current I flows for a time t is given by

$$Q = It \quad \text{or} \quad I = Q/t, \tag{1.1}$$

where I is in amperes (A), Q is in coulombs (C), and t is in seconds (s).

The question relates to the circuit shown in Fig. 1.2, where the number of electrons arriving at any point in any time interval

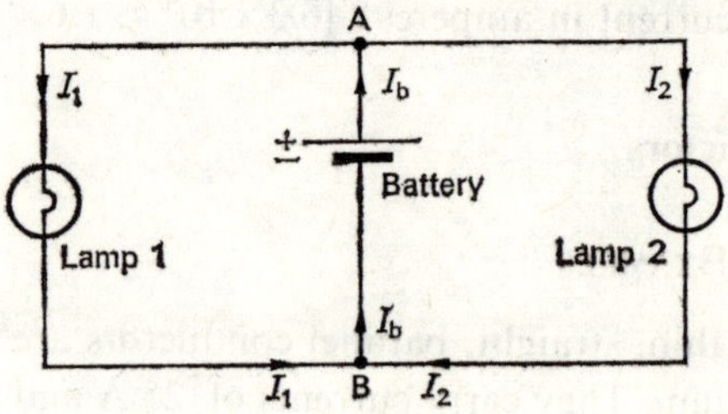

FIG. 1.2. Ex. 1.1.

equals the number leaving in the same interval. No means of storing an accumulation of electrons is to be considered in this circuit. Thus at either terminal A or B,

$$I_b = I_1 + I_2, \tag{1.2}$$

where I_b is the battery current and I_1 and I_2 are the lamp currents.

Solution

Using eqn. (1.1), the current through the first lamp

$$I_1 = Q/t = 7.5 \times 10^{18} \times 1.6 \times 10^{-19}/1 = 1.2 \text{ A}.$$

Then, from eqn. (1.2),

$$I_2 = I_b - I_1 = 1.6 - 1.2 = 0.4 \text{ A}. \tag{a}$$

Now from eqn. (1.1) the charge passed by I_2 in one second $=$ 0.4×1 C.

But each electron is a charge of 1.6×10^{-19} C. Therefore the rate of flow of electrons in I_2 is

$$0.4/(1.6 \times 10^{-19}) = 2.5 \times 10^{18} \text{ electrons/second.} \qquad \text{(b)}$$

1.2. Calculate the number of electrons which pass the terminals of an electric heater which draws a continuous current of 8.0 A for a period of 12 h. [2.2×10^{24}.]

1.3. How long would it take to supply a charge of one microcoulomb at the rate of one million electrons per millisecond? What would be the current in amperes? [6.3×10^3 s; 1.6×10^{-10} A.]

Parallel Conductors

1.4. Worked Example

Two long, thin, straight, parallel conductors are spaced 6.0 cm apart in a vacuum. They carry currents of 25 A and 30 mA respectively. Calculate (a) the force per unit length exerted on each conductor, and (b) the flux density of the magnetic field across the axis of each conductor.

Introduction

When an electric current flows in a wire, a magnetic field is produced in the surrounding space or medium. Evidence of this field is provided by the force exerted on a magnetic compass or on another conductor carrying a current, as in ex. 1.1. Such a force may be attributed to the existence of magnetic flux, which is imagined to pass around the current causing the field, and to have the same direction as a compass needle.

With two conductors as in the question, each current sets up a field of magnetic flux around it, and thereby exerts a force on the other conductor. In accordance with Newton's law of action and reaction, the forces on the two conductors are equal and opposite. Hence, in the following analysis, the suffixes 1 and 2 could be interchanged.

In Fig. 1.3 the loops show some of the directions of the magnetic flux produced by current I_1 and exerting the force F_2 on current I_2. The force is perpendicular to both the flux and the current where they cross.

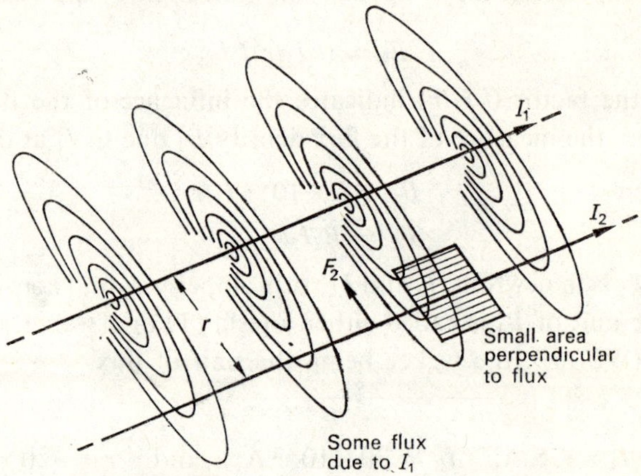

FIG. 1.3. Magnetic field of one conductor crossing another.

The force per unit length, F_2/l_2 or F_1/l_1, is found to be proportional to both currents I_1 and I_2 and inversely proportional to the distance r between the conductors. Thus

$$F_1/l_1 = F_2/l_2 = kI_1I_2/r, \tag{1.3}$$

where k is a constant which depends on the system of units and the medium in which the field is set up.

In M.K.S. units, the ampere (A) is defined as that value of both I_1 and I_2 which makes $F/l = 2\times10^{-7}$ newton per metre ($N\ m^{-1}$), when $r = 1$ metre (m) in a vacuum (see also ex. 1.1).

$$\therefore\ k = 2\times10^{-7}\ N\ A^{-2} \tag{1.4}$$

in vacuum.

The strength of the magnetic field in any small region is represented by the flux density B, i.e. the ratio of the quantity of flux passing through the region to the cross-section area through which it passes perpendicularly (see Fig. 1.3). Since F decreases with increase of r, so also must the flux density be less at greater distances from the current which causes it. This is represented roughly by the spacing of the loops of flux in Fig. 1.3.

An expression for B appears on rearranging eqn. (1.3):

$$F_2 = (kI_1/r)I_2l_2,$$

where the factor (kI_1/r) indicates the influence of the flux and is therefore the measure of the flux density B_1 due to I_1 at distance r.

Then $$B_1 = 2\times 10^{-7}\, I_1/r \tag{1.5}$$

and $$F_2 = B_1I_2l_2, \tag{1.6}$$

where F_2 is in newtons, I_1 and I_2 are in amperes, l_2 and r are in metres, and the unit of B_1 is called either a tesla (T) or a weber per square metre (Wb/m^2), the weber being the unit of flux.

Solution

(a) $I_1 = 25$ A, $I_2 = 30\times 10^{-3}$ A, and $r = 6.0\times 10^{-2}$ m.

Then eqns. (1.3) and (1.4) give the force per unit length:

$$F_1/l_1 = F_2/l_2 = 2\times 10^{-7}\times 25\times 30\times 10^{-3}/(6.0\times 10^{-2})$$
$$= 2.5\times 10^{-6}\text{ N m}^{-1}.$$

(b) Across I_2, the flux density due to I_1 is given by eqn. (1.5):

$$B_1 = 2\times 10^{-7}\, I_1/r = 2\times 10^{-7}\times 25/(6.0\times 10^{-2})$$
$$= 8.3\times 10^{-5}\text{ T}.$$

Similarly, across I_1,

$$B_2 = 2\times 10^{-7}\times 30\times 10^{-3}/(6.0\times 10^{-2})$$
$$= 1.0\times 10^{-7}\text{ T}.$$

1.5. Consider two conductors as in ex. 1.4 but with unknown currents and spacing. One current is then trebled and the other is halved. Calculate the change of spacing required to obtain a quarter of the original force. [+500%.]

1.6. A long, straight wire, carrying a current of 50 A, experiences a force per unit length of 2×10^{-2} N m^{-1} due to the magnetic field of a second wire running parallel to it at a distance of 5.0 cm in a vacuum. Calculate (a) the flux density of the magnetic field acting on the first wire, and (b) the current in the second wire. [(a) 4×10^{-4} T; (b) 100 A.]

Ammeters

1.7. Worked Example

A moving-coil instrument has a coil 1.5 cm square, with $20\frac{1}{2}$ turns of wire, pivoted in a magnetic field of flux density 0.20 T. The control springs have a total stiffness factor of 4.0×10^{-6} Nm per degree. Calculate the current which would give a deflection of 45°.

Introduction

The coil of a moving-coil instrument is pivoted about an axis, as in Fig. 1.4, so that it may turn under the influence of the forces acting on the sides of the coil which run across the curved faces of the pole pieces.

The force on each wire on each side is given by eqn. (1.6), i.e. $F = BIl$, where l is the length of one coil side, I is the current, and B is the density of the radial magnetic flux due to the poles. F is perpendicular to both l and B and therefore perpendicular to the plane of the coil. Also F acts on each side to produce rotation in the same direction.

If the coil has N turns, the total turning moment on each side is NFr, where r is the radius of the circle in which the coil can rotate. The resultant turning moment or torque T for both sides together is then given by

$$T = 2NFr = 2N(BIl)r.$$

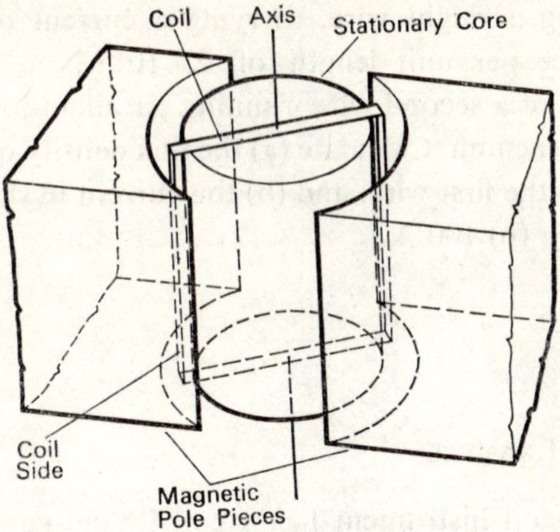

FIG. 1.4. Coil, core, and poles of a moving-coil instrument.

Now $2lr$ is the area A of the coil.

$$\therefore\ T = NBIA, \tag{1.7}$$

where T is in newton metres, B is in teslas, I is in amperes, and A is in square metres.

When the coil turns under the influence of this torque, the control springs are twisted and oppose the movement with a torque given by the stiffness factor S. The two torques balance in the equilibrium position.

$$\therefore\ NBIA = S\theta, \tag{1.8}$$

where S is in newton metres per degree and θ is the angular deflection in degrees.

Solution

I is found by rearranging eqn. (1.8): $I = S\theta/(NBA)$.

$$\therefore\ I = 4.0\times 10^{-6}\times 45/(20.5\times 0.20\times 1.5^2\times 10^{-4}) = 0.20\text{ A}.$$

1.8. A moving-coil instrument has a coil of area 2.0 cm² pivoted in a magnetic field of flux density 0.20 T. The control springs together have a stiffness factor of 5.0×10^{-6} Nm per degree. Calculate the number of turns required to obtain a deflection of 60° when the current is 80 mA. [$93\frac{1}{2}$].

1.9. New control springs and a rewound coil are fitted to a moving-coil instrument which originally had a sensitivity of 10 cm deflection per milliamp. Calculate the new sensitivity if the new springs have a stiffness factor 30% greater than the old and the number of turns of the coil has been decreased from $80\frac{1}{2}$ to $75\frac{1}{2}$. [7.2 cm mA^{-1}.]

E.M.F and P.D.

1.10. Worked Example

Two lamps are connected in series across a supply of e.m.f. 240 V and take a current of 0.750 A. One lamp takes energy of 9.00×10^4 J in a period of 20.0 min. Calculate (a) the energy taken by the other lamp in the same time, and (b) the p.d. across each lamp.

Introduction

A source of electrical energy has an electromotive force (e.m.f.) of 1 volt if it expends energy of 1 joule in passing a charge of 1 coulomb in a current around the circuit. Hence if a source expends energy W on the passage of charge Q, the e.m.f.

$$E = W/Q, \tag{1.9}$$

where E is in volts (V), W in joules (J), and Q in coulombs (C).

Any two points of the circuit have a potential difference (p.d.) of 1 volt if the energy provided by the current between these points is 1 joule when a charge of 1 coulomb passes. Hence if energy W leaves the circuit when a charge Q passes, the p.d.

$$V = W/Q, \tag{1.10}$$

where V is in volts (V), W in joules (J), and Q in coulombs (C).

Solution

The quantity of charge which passes all points of the circuit when the current of 0.75 A flows for a time of 20 min is

$$Q = It = 0.75\times(20\times60) = 900 \text{ C}.$$

Then from eqn. (1.9) the energy given by the e.m.f. in this time is

$$W = EQ = 240\times900 = 2.16\times10^5 \text{ J}.$$

But one lamp takes energy

$$W_1 = 9\times10^4 \text{ J}.$$

Therefore the other lamp takes energy

$$W_2 = (2.16\times10^5)-(9\times10^4) \text{ J}$$
$$= 1.26\times10^5 \text{ J}. \quad \text{(a)}$$

Next, from eqn. (1.10) the p.d. across the first lamp is

$$V_1 = W_1/Q = 9\times10^4/900 = 100 \text{ V}. \quad \text{(b)}$$

Similarly, the p.d. across the other lamp through which the same charge passes is

$$V_2 = W_2/Q = 1.26\times10^5/900 = 140 \text{ V}. \quad \text{(b)}$$

As a partial check, note that

$$V_1+V_2 = W_1/Q+W_2/Q = \text{total } W/Q = E.$$

Thus the sum of p.d.s in series equals the e.m.f.; i.e. 100+140 = 240 V.

1.11. Calculate the e.m.f. acting in a circuit which takes energy of 240 J from the source when a current of 0.5 A flows for 2 min. [4 V.]

1.12. The energy dissipated by a current of 15 A flowing for a time of 30 min through a heater is 6.5×10^6 J. Calculate the p.d. between the heater terminals. [240 V.]

1.13. An e.m.f. of 25 V drives a current of 0.35 A through a lamp filament. Calculate the energy time and required to pass a charge of 1000 C. [2.5×10^4 J; 48 min.]

1.14. An electric heater has two elements each of which produces 10^6 J of heat in a period of 30 min when carrying a current of 4.0 A. Calculate the e.m.f. and current to be supplied by a single source when the elements are connected (a) in series, and (b) in parallel. [(a) 280 V, 4 A; (b) 140 V, 8 A.]

Voltmeters

1.15. Worked Example

A moving-coil voltmeter takes a current of 10 mA when its terminal p.d. is 5.0 V. When connected in parallel with a lamp and a battery it indicates a p.d. of 15 V between the lamp terminals. The lamp dissipates energy of light and heat at the rate of 300 J min^{-1}. Calculate (a) the lamp current, and (b) the battery current.

Introduction

Ohm's law states that, at constant temperature, the current in a conductor between two points is proportional to the p.d. between these points. In particular, the coil and internal connections of a moving-coil instrument form a conductor. Thus the current through the instrument is proportional to the p.d. between its terminals.

Now, as shown in ex. 1.7, the deflection θ is proportional to the current I. Therefore the deflection is proportional also to the terminal p.d. V, and the instrument scale may be calibrated to indicate V directly. Thus a moving-coil instrument may indicate either the current through it or the p.d. between its terminals when calibrated as a voltmeter or millivoltmeter.

Solution

(a) The lamp current in amperes $I_L = Q/t$ and $Q = W/V$, where Q is the charge in coulombs passing in time t in seconds, and W is the energy in joules dissipated as the charge Q passes between the lamp terminals at a p.d. of V in volts.

$$\therefore\ I_L = W/(Vt) = 300/(15\times 60) = \tfrac{1}{3}\text{ A}$$
$$= 330 \text{ mA.} \qquad \text{(a)}$$

(b) With the voltmeter, lamp, and battery all in parallel, the p.d. of 15 V applies equally to each of them. Now the instrument current is 10 mA for a p.d. of 5.0 V, and therefore, by the proportionality of Ohm's law, it is 30 mA for a p.d. of 15 V.

The battery supplies both lamp current and voltmeter current.

$$\therefore\ \text{battery current} = 330+30 = 360 \text{ mA.} \qquad \text{(b)}$$

1.16. A voltmeter is connected in parallel with a heater element which carries a current of 4.0 A. The voltmeter is known to require a current of 1.0 mA at a p.d. of 1.0 V for a deflection of 1.0°. Calculate the energy dissipated per hour by the heater, when the voltmeter deflection is 60°. [8.6×10^5 J.]

1.17. A particular ammeter requires a current of 0.50 A for full-scale deflection of 120°, and a particular voltmeter requires a current of 40 mA at a p.d. of 100 V for the same full-scale deflection. When connected in the circuit shown in Fig. 1.5, the deflection of the ammeter is 60° and that of the voltmeter is 90°. Calculate (a) the current in the resistor, and (b) the energy dissipated in one minute by the resistor. [(a) 220 mA; (b) 990 J.]

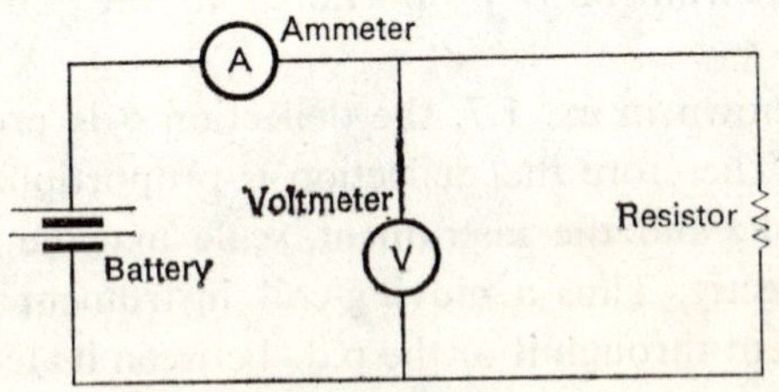

FIG. 1.5. Ex. 1.17.

CHAPTER 2

RESISTANCE

P.D. and Current

2.1. Worked Example

When a resistor carries a current of 2.5×10^{16} electrons per second its terminals have a p.d. of 2 kV. Calculate the resistance of the resistor.

Introduction and Solution

The resistance R of a conductor or resistor is defined as the ratio of the p.d. V between its ends to the current I passing through it.

Thus $R = V/I$. If V is measured in volts and I in amperes, the unit of R is the ohm (Ω).

Now the current is given as 2.5×10^{16} electrons/second and, from the reference tables, the charge of one electron is 1.6×10^{-19} C.

$$\therefore\ I = 2.5\times10^{16}\times1.6\times10^{-19}\ \text{C s}^{-1} \text{ or A}$$
$$= 4\times10^{-3}\ \text{A}.$$

Also the p.d. is given as 2 kV. $\quad \therefore\ V = 2\times10^{3}$ V.

Then $\quad R = V/I = 2\times10^{3}/(4\times10^{-3}) = 0.5\times10^{6}\ \Omega = \frac{1}{2}\ \text{M}\Omega.$

2.2. Calculate the unspecified current, p.d., or resistance in each of the following cases where two of the three quantities are given in abbreviated form: (a) 20 V, 5000 Ω; (b) 50 μA, 22 kΩ; (c) 40 V,

10 mA; (d) 5 MΩ, 3 mA; (e) 20 kΩ, 50 mV; (f) 5 μA, 2.5 kV. [(a) 4 mA; (b) 1.1 V; (c) 4 kΩ; (d) 15 kV; (e) 2.5 μA; (f) 500 MΩ.]

2.3. A lamp running on a supply of e.m.f. 240 V has a resistance of 500 Ω. Calculate the current in amperes and in electrons per hour. [0.48 A; 1.08×10^{22} e h^{-1}.]

2.4. A p.d. of 25 V is established across a resistor of $\frac{1}{2}$ MΩ. Calculate the charge which passes in $\frac{1}{2}$ h. [0.09 C.]

2.5. A circuit is connected to a battery of e.m.f. 6.0 V for a time of 10 min. What is the resistance of the circuit if it draws (a) a charge of 0.10 C, and (b) energy of 0.10 J? [(a) 36 kΩ; (b) 0.22MΩ.]

Resistors in Series

2.6. Worked Example

A lamp has a rating of 12.0 V, 0.300 A. What resistor is needed in series to run the lamp correctly from a supply of 50.0 V?

Introduction and Solution

At the end of the solution of ex. 1.10 it was shown that conductors connected in series share the total p.d. or e.m.f.; i.e. for two conductors in series,

$$V_1 + V_2 = E. \tag{2.1}$$

Thus if the lamp p.d. $V_1 = 12$ V when the circuit e.m.f. $E = 50$ V, then the resistor p.d. $V_2 = 50 - 12 = 38$ V.

Now the lamp and resistor in series must pass the same current $I = 0.3$ A. Therefore the required resistance

$$R_2 = V_2/I = 38/0.3 = 127\ \Omega.$$

Alternatively, dividing eqn. (2.1) by the current, which is the same throughout the series circuit, $(V_1/I)+(V_2/I) = (E/I)$, i.e.

$$R_1 + R_2 = R, \tag{2.2}$$

where R_1 and R_2 are the resistances of the conductors in series and R is the total circuit resistance.

Then for the lamp,

$$R_1 = V_1/I = 12/0.3 = 40\ \Omega,$$

and for the complete circuit,

$$R = E/I = 50/0.3 = 167\ \Omega.$$

So from eqn. (2.2) the required resistance

$$R_2 = R - R_1 = 167 - 40 = 127\ \Omega.$$

2.7. A radio set has four valves with heaters connected in series. Each heater requires a current of 200 mA at a p.d. of 4.0 V. Calculate the series resistor required to run from a 24 V supply. [40 Ω.]

2.8. Two resistors of resistance 2.0 Ω and 6.0 Ω respectively are connected in series across a supply of e.m.f. 24 V. Calculate the p.d. across each resistor. [6.0 V; 18 V.]

2.9. A pair of leads, each of resistance 0.10 Ω, supplies an electric motor from a generator giving an output of 500 V, 23 A. Calculate the p.d. at the motor terminals. [495.4 V.]

2.10. A battery of accumulators drives a current of 4.0 A through a resistor. When another resistor of 3.0 Ω is connected in series with the first, the current is 3.0 A. Calculate the e.m.f. of the battery and the resistance of the first resistor. [36 V; 9 Ω.]

2.11. A lamp requiring a p.d. of 12 V and a current of 4.0 A is to be connected to a battery of e.m.f. 15 V. Each connecting lead has a resistance per unit length of 5.0 Ω km^{-1}. Calculate the maximum distance at which the lamp may be operated correctly. [75 m.]

2.12. An ammeter is in series with a resistor of 240 Ω and a battery of e.m.f. 12 V. The resistance of the ammeter is 10 Ω. Calculate the error in assuming that the current is unaffected by removing the meter and reclosing the circuit. [4%.]

2.13. A circuit has a resistance of 300 Ω and a constant e.m.f. What is the greatest resistance of an ammeter which can be inserted to indicate the original current with an error of less than 3%? [9.28 Ω.]

Internal Resistance

2.14. Worked Example

Calculate the maximum current obtainable from a battery of constant e.m.f. and internal resistance if a load of 4.8 Ω draws a current of 2.0 A and a load of 9.8 Ω draws a current of 1.0 A.

Introduction

The battery may be represented by an e.m.f. E acting in series with its internal resistance r (Fig. 2.1).

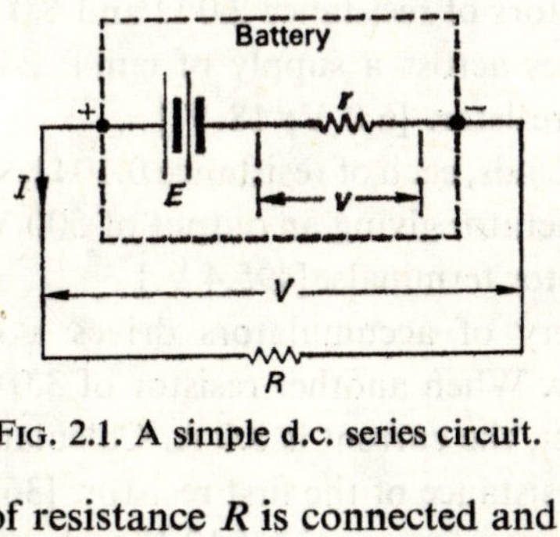

FIG. 2.1. A simple d.c. series circuit.

When a load of resistance R is connected and a current I flows through the circuit, the e.m.f. equals the total potential drop around the circuit, i.e.

$$E = V+v, \tag{2.3}$$

where V = battery terminal p.d. = p.d. of load = IR, and v = p.d. "lost" in battery = Ir.

Therefore, from eqn. (2.3),

$$IR = E-Ir, \tag{2.4}$$

i.e. terminal or load p.d. = e.m.f. − lost p.d.

Furthermore, from eqn. (2.4),

$$I = E/(R+r).$$

Therefore I is a maximum when $R = 0$, i.e. when the circuit resistance $(R+r)$ is a minimum.

Solution

In the two cases given, eqn. (2.4) gives

$$2\times4.8 = E-2r \qquad (2.5)$$

and

$$1\times9.8 = E-1r. \qquad (2.6)$$

Now eqn. (2.6) minus eqn. (2.5) gives

$$9.8-9.6 = r = 0.2\ \Omega.$$

Then substituting in eqn. (2.6),

$$E = 9.8+0.2 = 10\ \text{V}.$$

$$\therefore\ \text{maximum } I = E/r = 10/0.2 = 50\ \text{A}.$$

2.15. Calculate the terminal p.d. of a battery of e.m.f. 24.0 V with internal resistance 1.5 Ω when the battery is connected in series with resistors of 3.7 Ω and 2.8 Ω. [19.5 V.]

2.16. Calculate the current and resistance of a 3.5 volt lamp which is running correctly from a battery of internal resistance 2 Ω and e.m.f. 4.5 V. [$\frac{1}{2}$ A; 7 Ω.]

2.17. A battery gives a current of 5 A on a certain load. If the load is increased by 3 Ω the current drops to 4 A, and if the load is shorted the current rises to 30 A. Calculate the e.m.f. and internal resistance of the battery. [60 V; 2 Ω.]

2.18. How many cells, each of internal resistance 2 Ω and e.m.f. 1.5 V, are needed to supply a current of 250 mA when connected in series with a load of resistance 32 Ω? [8.]

2.19. Six accumulators are to be charged in series at a current of 20 A. Each gives a back e.m.f. of 2.4 V and has internal resistance of 0.03 Ω. Calculate the series ballast resistance needed with a charger of e.m.f. 48 V and internal resistance 0.5 Ω. [1 Ω.]

Resistors in Parallel

2.20. Worked Example

Three resistors of resistance 20 Ω, 30 Ω, and 40 Ω respectively are connected in parallel across a battery and draw a total current of 2.6 A. Calculate (a) the equivalent single resistance, and (b) the current in each resistor.

Introduction

Since the resistors 1, 2, 3 are connected in parallel (Fig. 2.2), the sum of their currents equals the battery current I (see also ex. 1.1), i.e.

$$I = I_1 + I_2 + I_3. \tag{2.7}$$

Assuming that the leads and connections are of negligible resistance, each of the parallel resistors has the same p.d. V as the battery terminals.

$$\therefore\ I_1 = V/R_1, \quad I_2 = V/R_2, \quad I_3 = V/R_3.$$

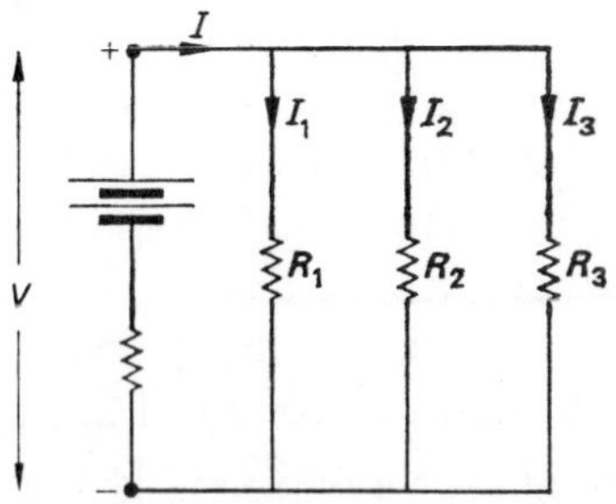

FIG. 2.2. A simple d.c. parallel circuit.

The equivalent single resistor of resistance R would draw the same battery current I at the same terminal p.d. V.

$$\therefore\ I = V/R.$$

Then, substituting in eqn. (2.7),

$$V/R = V/R_1 + V/R_2 + V/R_3.$$

$$\therefore\ 1/R = 1/R_1 + 1/R_2 + 1/R_3. \qquad (2.8)$$

Solution

(a) Substituting the values of resistance given in the problem,

$$1/R = 1/20 + 1/30 + 1/40 = (6+4+3)/120 = 13/120.$$

Therefore the equivalent single resistor has resistance

$$R = 120/13 = 9.2\ \Omega. \qquad \text{(a)}$$

(b) Since $V = IR$,

$$V = 2.6 \times 120/13 = 24\ \text{V}.$$

Therefore the current divides so that

$$\left.\begin{aligned} I_1 &= V/R_1 = 24/20 = 1.2\ \text{A}, \\ I_2 &= V/R_2 = 24/30 = 0.8\ \text{A}, \\ I_3 &= V/R_3 = 24/40 = 0.6\ \text{A}. \end{aligned}\right\} \qquad \text{(b)}$$

2.21. Two resistors connected in parallel draw a current of 5.0 A from a battery of negligible internal resistance. One resistor is of 10 Ω resistance, and the other draws a current of 2.0 A. Calculate the resistance of the second resistor. [15 Ω.]

2.22. Calculate the resistance of a stranded conductor which has 13 strands each of resistance 0.65 Ω. [0.05 Ω.]

2.23. What resistor in parallel with one of resistance 5.050 Ω is needed to give a resultant resistance of exactly 5.000 Ω? [505 Ω.]

2.24. Three resistors connected in parallel make a resultant resistance of 10 Ω. Two of the resistors have resistances of 20 Ω and 40 Ω respectively. Calculate the value of the third resistance. [40 Ω.]

2.25. A resistor of 5.0 Ω is connected across a battery of constant terminal p.d. Calculate the resistance of a resistor which may be connected in series or in parallel so that the battery current when this resistor is in parallel is five times greater than when it is in series. [13 Ω or 1.9 Ω.]

Shunts and Multipliers

2.26. Worked Example

The current for full-scale deflection of a milliammeter is changed trom 10.0 mA to 5.00 A by a shunt of resistance 0.0200 Ω. Calculate fhe multiplier resistance required to convert the instrument without the shunt to a voltmeter indicating 100 V at full-scale deflection.

Introduction

As explained in ex. 1.15, a moving-coil instrument may be used either as an ammeter or as a voltmeter, since the p.d. v between its terminals is proportional to the current i which operates it. Thus $v/i = r$, where r is the resistance of the instrument.

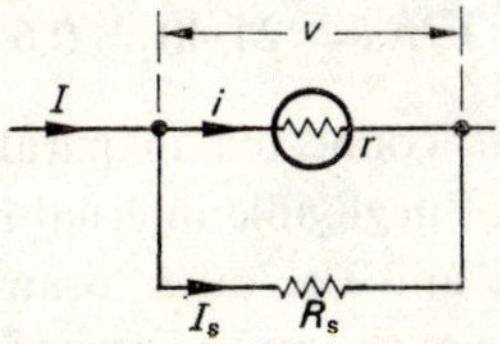

FIG. 2.3. A shunt resistance to increase the range of an ammeter.

To measure a current I higher than the normal value for full-scale deflection (f.s.d.), a shunt of resistance R_S may be connected across the terminals, inside or outside the instrument case (Fig. 2.3). The shunt diverts the excess current $I_S = (I-i)$, while the operating current i passes through the coil of the instrument. I may then be calculated from the indicated value or the scale may be numbered to indicate I directly.

Now the terminal p.d. v is common to both the coil and the shunt.

$$\therefore\ i = v/r \quad \text{and} \quad I_S = v/R_S.$$

Then the factor S by which i must be multiplied to obtain I is given by

$$S = \frac{I}{i} = \frac{I_S+i}{i} = \frac{v/R_S+v/r}{v/r} = \frac{r}{R_S}+1. \tag{2.9}$$

So, if R_S is much less than r, the current range is much extended.

To measure a p.d. V higher than the normal value for f.s.d., a "multiplier" of resistance R_m may be connected in series with the instrument. The multiplier then takes the excess p.d. $V_m = V-v$, while the instrument carries current i at p.d. v (see Fig. 2.4). V may be calculated from the indicated value v, or the scale may be numbered to indicate V directly.

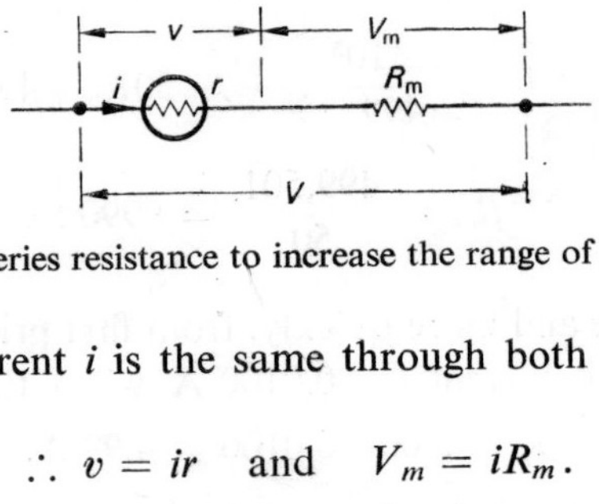

FIG. 2.4. A series resistance to increase the range of a voltmeter.

Now the current i is the same through both the coil and the multiplier.

$$\therefore\ v = ir \quad \text{and} \quad V_m = iR_m.$$

The factor M by which v must be multiplied to obtain V is given by

$$M = \frac{V}{v} = \frac{V_m+v}{v} = \frac{iR_m+ir}{ir} = \frac{R_m}{r}+1. \tag{2.10}$$

Thus, if R_m is much greater than r, the voltage range is much extended.

Solution

In the question the shunt factor

$$S = \frac{I}{i} = \frac{5.00}{10.0 \times 10^{-3}} = 500$$

and, from eqn. (2.9),

$$S = \frac{r}{R_S} + 1 = \frac{r}{0.0200} + 1 = 50r + 1.$$

$$\therefore\ 50r + 1 = 500. \quad \therefore\ r = \frac{499}{50}\ \Omega.$$

Then, with the multiplier, the factor $M = V/v = V/ir$

$$= \frac{100}{100 \times 10^{-3} \times 499/50} = \frac{5 \times 10^5}{499}.$$

Also, from eqn. (2.10),

$$M = \frac{R_m}{r} + 1 = \frac{R_m}{(499/50)} + 1.$$

$$\therefore\ \frac{R_m}{(499/50)} + 1 = \frac{5 \times 10^5}{499}. \quad \therefore\ 50R_m + 499 = 500{,}000.$$

$$\therefore\ R_m = \frac{499{,}501}{50} = 9990\ \Omega.$$

Alternatively, and more quickly, from first principles: with the shunt, the coil current $i = 0.0100$ A when the shunt current

$$I_s = 5.00 - 0.0100 = 4.99\ \text{A}.$$

$$\therefore\ \text{the common p.d. for f.s.d.} = ir = I_s R_s$$
$$= 4.99 \times 0.0200 = 0.0998\ \text{V}.$$

For the voltmeter arrangement, the multiplier must have a p.d. of $100 - 0.0998 = 99.9002$ V, and must carry the current $i = 0.0100$ A.

$$\therefore\ R_m = \frac{99.9002}{0.010} = 9990\ \Omega.$$

2.27. An instrument of resistance 50 Ω takes a current of 1.0 mA for f.s.d. Calculate (a) the p.d. for f.s.d. when using a multiplier of 10 kΩ, and (b) the current for f.s.d. when using a shunt of 1.0 Ω. [(a) 10 V; (b) 51 mA.]

2.28. The p.d. for f.s.d. of a voltmeter is changed from 50 mV to 50 V by a multiplier of resistance 4000 Ω. Calculate the shunt resistance required to convert the instrument, without the multiplier, to an ammeter requiring a current of 10 A for f.s.d. [0.0050 Ω.]

2.29. A moving-coil instrument requires a current of 0.015 A and a p.d. of 0.075 V for f.s.d. Calculate the shunt and multiplier resistances required to extend the range to f.s.d. values of (a) 15 V, and (b) 3.0 A. [(a) 1.0 kΩ; (b) 0.025 Ω.]

2.30. When an instrument is fitted with a shunt of resistance 1.00 Ω, the current range is multiplied by 100. Calculate the multiplier resistance required to multiply the voltage range by 100. [9.8 kΩ.]

CHAPTER 3

NETWORKS

Series–Parallel Circuits

3.1. Worked Example

Calculate the current in the 3 Ω resistor of the circuit shown in Fig. 3.1.

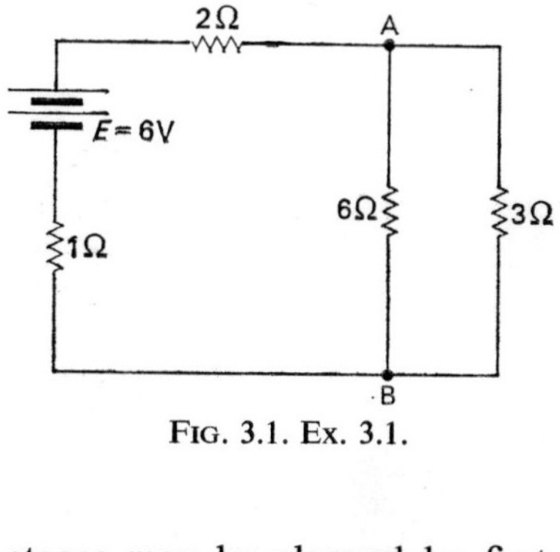

FIG. 3.1. Ex. 3.1.

Introduction

The various stages may be planned by first stating the final stage, and then considering what is required of each preceding stage in the reverse order.

Thus the final stage:

(1) Required current I = (p.d. AB) ÷ (3 Ω).
(2) (p.d. AB) = E − (p.d. across 2 Ω and 1 Ω in series).
(3) (p.d. across 2 Ω and 1 Ω in series) = (battery current) × (2 Ω + 1 Ω).

(4) Battery current = $E \div$ (resultant resistance of circuit).
(5) Resultant resistance of circuit = $1\,\Omega + 2\,\Omega +$ (resultant of $3\,\Omega$ and $6\,\Omega$ in parallel).
(6) Resultant of $3\,\Omega$ and $6\,\Omega$ in parallel may be calculated readily.

Solution

Working through these stages from (6) to (1):

(6) $1/R = \frac{1}{3} + \frac{1}{6} = \frac{1}{2}$. $\quad \therefore R = 2\,\Omega$.
(5) Circuit resistance = $1+2+2 = 5\,\Omega$.
(4) Battery current = $\frac{6}{5} = 1.2$ A.
(3) P.D. across $2\,\Omega$ and $1\,\Omega$ in series = $1.2\times 3 = 3.6$ V.
(2) P.D. $AB = 6-3.6 = 2.4$ V.
(1) Current through $3\,\Omega$ resistor $= \dfrac{2.4}{3} = 0.8$ A.

3.2. Worked Example

A galvanometer of resistance $40\,\Omega$ is connected in series with a resistor of $60\,\Omega$ to a supply of constant e.m.f. and negligible internal resistance. Calculate the resistance required as a shunt across the galvanometer to halve the indicated current.

Solution

Assuming correct operation, the galvanometer indicates the current through it.

Thus, without the shunt, the indicated current I_1 is that through the $40\,\Omega$ galvanometer and $60\,\Omega$ resistor in series (Fig. 3.2).

$$\therefore\ I_1 = \frac{E}{40+60} = 0.01\times E.$$

With the shunt of unknown resistance X, the resultant resistance R between the galvanometer terminals is given by

$$\frac{1}{R} = \frac{1}{40} + \frac{1}{X} = \frac{X+40}{40X}. \qquad \therefore\ R = \frac{40X}{X+40}.$$

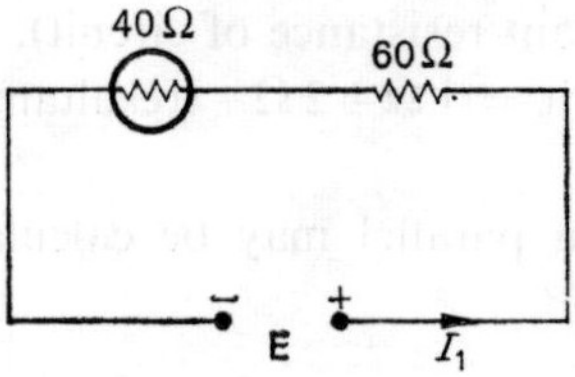

FIG. 3.2. Ex. 3.2 — without galvanometer shunt.

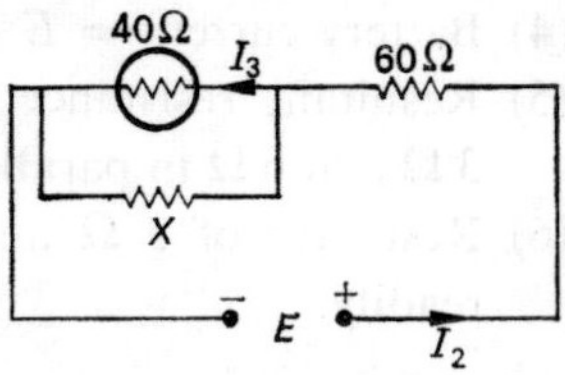

FIG. 3.3. Ex. 3.2 — with galvanometer shunt.

Therefore in Fig. 3.3 the resultant circuit resistance

$$= 60+\left(\frac{40X}{X+40}\right)$$

and total current

$$I_2 = \frac{E}{60+[40X/(X+40)]} = \frac{E(X+40)}{100X+2400}.$$

Now p.d. V of galvanometer terminals

$$= I_2 \times R = \frac{E(X+40)}{100X+2400} \times \frac{40X}{(X+40)} = \frac{40EX}{100X+2400}.$$

Therefore galvanometer current

$$I_3 = \frac{V}{40} = \frac{EX}{100X+2400}.$$

Then, since I_3 must be $\frac{1}{2} \times I_1$,

$$\frac{EX}{100X+2400} = \frac{1}{2} \times 0.01 \times E.$$

$$\therefore\ X = 0.5X+12. \qquad \therefore\ X = 24\ \Omega.$$

3.3. Calculate the different total values of resistance which may be obtained by taking three of the four following resistors in various ways and interconnecting them: 2Ω, 2Ω, 4Ω, 4Ω. [$\frac{4}{5}$, 1, $1\frac{1}{2}$, $1\frac{3}{5}$, 2, $2\frac{2}{5}$, $3\frac{1}{3}$, 4, 5, $5\frac{1}{3}$, 8, 10 Ω.]

3.4. Four resistors, each of 5 Ω, are connected as the sides of a square. A current of 2 A is drawn when a battery of constant terminal p.d. is connected across diametrically opposite corners. Calculate the current drawn (a) when one resistor is changed to 15 Ω, and (b) when the battery is reconnected across one pair of adjacent corners of the square of 5 Ω resistors. [(a) $1\frac{1}{2}$ A; (b) $2\frac{2}{3}$ A.]

3.5. A battery of e.m.f. 6 V and negligible internal resistance is connected in series with two resistors each of resistance 10 kΩ. It is required to halve the current in one of these resistors by connecting another resistor in parallel with it. Calculate the resistance required. [5 kΩ.]

3.6. A battery of negligible internal resistance is connected in series with two resistors each of resistance 100 Ω. A voltmeter indicates 6.00 V when connected across both resistors, and 2.90 V when connected across each. Calculate the resistance of the voltmeter. [1450 Ω.]

3.7. In the circuit of Fig. 3.4 the ammeter *A* has a resistance of 4.0 Ω and indicates a current of 3.6 A. Calculate (a) the terminal p.d. of the battery *B*, and (b) the current in the 1.0 Ω resistor. [(a) 21 V; (b) 1.8 A.]

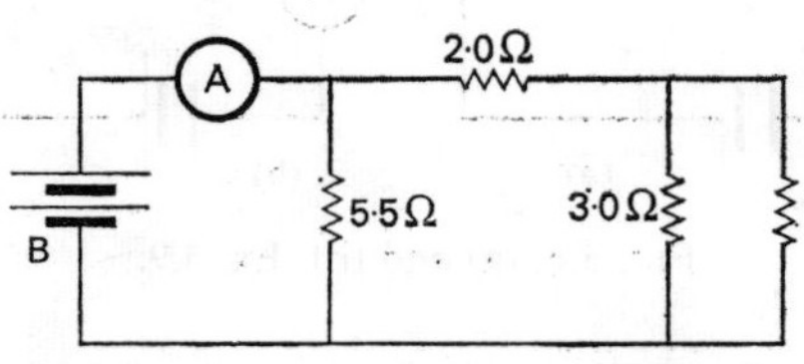

FIG. 3.4. Ex. 3.7.

3.8. A battery of e.m.f. 12 V and internal resistance 1 Ω is connected across the terminals *A* and *B* of the circuit shown in Fig. 3.5. Calculate the resistance *R* and the terminal p.d. of the battery, when *R* draws a current of $\frac{1}{3}$ A. [3 Ω; 11 V.]

3.9. An ammeter *A* and a voltmeter *V* are used to measure the unknown resistance *R* of a resistor. Connections are made as in

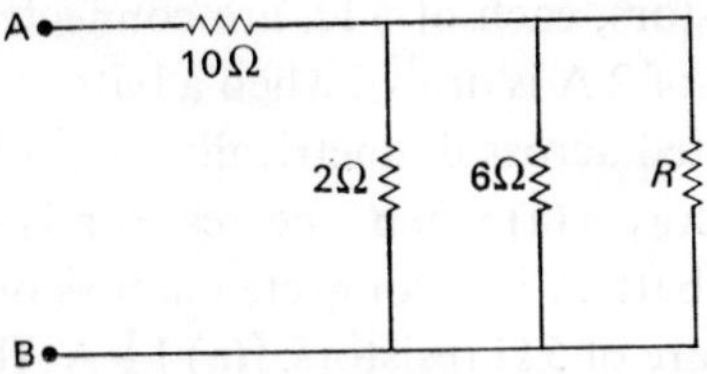

FIG. 3.5. Ex. 3.8.

Fig. 3.6(a) and (b), using the same battery of negligible internal resistance and constant e.m.f. in each case. The instrument readings are 15.0 mA and 48.0 V in case (a), and 20.0 mA and 47.5 V in case (b). Calculate the value of R and, in each case, the error in assuming that the simple ratio of instrument readings gives the true value of R. [3175 Ω; (a) +0.79%; (b) −25.2%.]

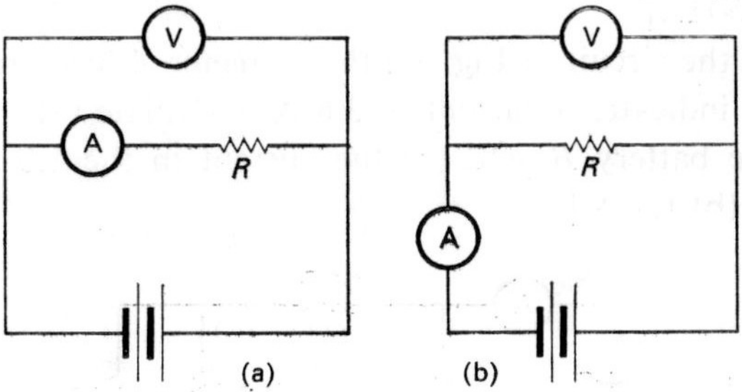

FIG. 3.6. (a) and (b). Ex. 3.9.

3.10. A battery of e.m.f. 24 V is in series with two resistors of resistance 10 kΩ and 20 kΩ respectively. Calculate the change of p.d. across the 20 kΩ resistor when a voltmeter is connected across it, and hence find the error in measuring the p.d. without regard to this change when the voltmeter resistance is (a) 2.5 kΩ, and (b) 25 kΩ V^{-1}. [(a) 73%; (b) 21%.]

Balanced Potentiometers

3.11. Worked Example

A slide-wire potentiometer has a wire of length 100 cm connected across a battery of e.m.f. 3 V. For a Weston standard cell of e.m.f. 1.0186 V, the point of balance is at a distance of 35.6 cm from one end *O* of the wire. Calculate (a) the unknown e.m.f. of a cell for which the point of balance is 55.0 cm from *O*, (b) the position of the point of balance for a cell of e.m.f. 1.45 V, and (c) the error in ignoring the standardization with the Weston cell and assuming that the supply battery produces a p.d. of 3.00 V between the ends of the wire.

Introduction

The potentiometer is balanced when the adjustable contact *P* (Fig. 3.7) on the wire *OQ* is set so that the galvanometer *G* indicates

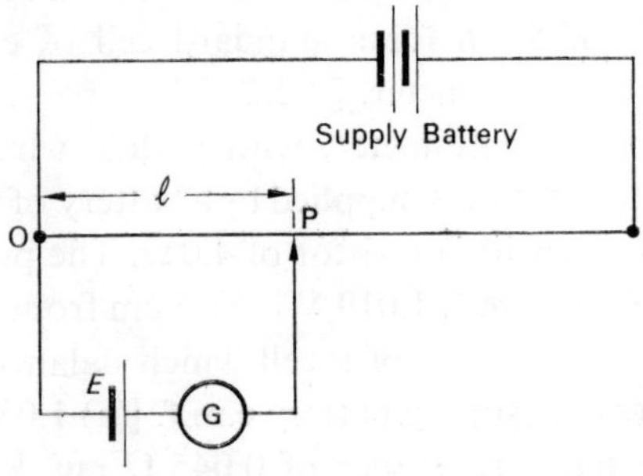

FIG. 3.7. Simple d.c. potentiometer (ex. 3.11).

zero current. This current is zero when the e.m.f. *E* is equal and opposite to the p.d. between *O* and *P* on the wire. Assuming a uniform wire and a constant p.d. between its ends, the potential gradient along the wire is uniform at balance.

Solution

The potential gradient is

$$\frac{\text{p.d. } OP}{\text{length } OP} = \frac{E}{l} = \frac{1.0186}{35.6} = 0.0286 \text{V cm}^{-1}.$$

(a) For balance at 55 cm, $E = 0.0286 \times 55 = 1.57$ V. (a)

(b) For $E = 1.45$ V, l is given by $0.0286 = 1.45/l$ V cm^{-1}.

$$\therefore\ l = \frac{1.45}{0.0286} = 50.7 \text{ cm.} \qquad \text{(b)}$$

(c) If it is assumed that the p.d. is 3.00 V across 100 cm of wire then the potential gradient is

$$\frac{3.00}{100} = 0.0300 \text{ V cm}^{-1}.$$

This differs from the value determined by standardization by $0.0300 - 0.0286 = 0.0014$ V cm^{-1}. Therefore

$$\text{the error} = \frac{0.0014}{0.0286} \times 100\% = 4.9\%. \qquad \text{(c)}$$

3.12. When the accuracy of a voltmeter is tested on a slide-wire potentiometer, balance is obtained at 66.0 cm for a meter indication of 1.50 V and at 45.5 cm for a standard cell of e.m.f. 1.018 V. Calculate the error of the meter. [+1.7%.]

3.13. A simple potentiometer, with a slide wire of resistance 3.98 Ω and length 99.5 cm, is supplied by a battery of internal resistance 0.02 Ω in series with a resistor of 4.0 Ω. The point of balance for a standard cell of e.m.f. 1.019 V is 81.5 cm from one end of the wire. Calculate (a) the e.m.f. of a cell which balances for a length of 84.0 cm, and (b) the supply battery e.m.f. [(a) 1.05 V; (b) 2.5 V.]

3.14. A wire has a resistance of 0.045 Ω cm^{-1}. Calculate the length required for a simple potentiometer indicating 2.0 mV mm^{-1} when supplied by a battery of e.m.f. 2.10 V and internal resistance 0.05 Ω. [104 cm.]

3.15. A potentiometer with a slide wire of length 100 cm and resistance 5.0 Ω gives a balance for a length of 48.5 cm with a Weston cell of e.m.f. 1.0186 V. Assuming the supply battery to have constant e.m.f. and negligible internal resistance, estimate the extra resistance required in series with the wire and battery to make the millimetre scale give a direct indication of the p.d. in millivolts. [5.5 Ω.]

3.16. The balance length on a potentiometer wire is 75.0 cm for a certain cell on no load and 73.6 cm for the same cell with a load resistor of 10 Ω across its terminals. Calculate the internal resistance of the cell. [0.19 Ω.]

3.17. The accuracy of a milliammeter is checked on a slide-wire potentiometer with a standard cell of e.m.f. 1.02 V and a standard resistor of resistance 20 Ω. Balance occurs at 43.0 cm for the cell and at 83.5 cm for the resistor carrying a current which the milliammeter indicates to be 100 mA. Calculate the error of the milliammeter. [+1.0%.]

3.18. Two potentiometer wires each 50 cm long are connected in series with a battery of e.m.f. 6.0 V and resistance 10 Ω. One wire *A* has resistance of 15 Ω and the other, *B*, 25 Ω. A p.d. of 3.0 V is to be tapped off the wires. Determine the position of one tapping point when the other is (a) the mid-point of *A*, and (b) the mid-point of *B*. [(a) 35 cm; (b) 42 cm from joint of *A* and *B*.]

3.19. A slide-wire potentiometer is balanced (a) at 47.9 cm for the e.m.f. 1.0186 V of a standard cell, (b) at 70.5 cm for the e.m.f. of a dry cell, and (c) at 22.0 cm for the dry cell and standard cell in series opposition. Calculate the end correction of the potentiometer and the e.m.f. of the dry cell. [0.6 cm; 1.49 V.]

Balanced Bridges

3.20. Worked Example

Figure 3.8 shows a balanced Wheatstone bridge circuit with a supply accumulator of negligible internal resistance and e.m.f. of 2 V. Calculate the resistance R_1 and the p.d. V_3 given that $R_2 =$ 200 Ω, $R_3 =$ 100 Ω, and $R_4 =$ 1000 Ω.

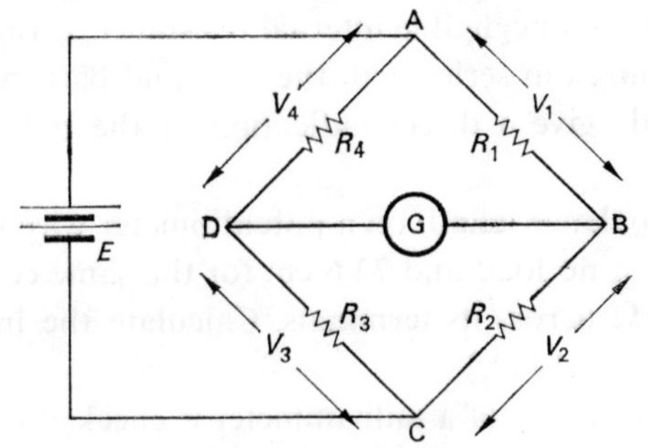

FIG. 3.8. Balanced Wheatstone bridge (exs. 3.20 and 3.21).

Introduction

The four resistors form two potential dividers across the supply battery, i.e. $V_1+V_2 = E$, and $V_4+V_3 = E$. The galvanometer G bridges from one point of division B to the other point D. The bridge is balanced when G indicates zero current from B to D.

R_1 and R_2 then carry the same current I_1, so that $V_1 = I_1R_1$ and $V_2 = I_1R_2$.

$$\therefore \frac{V_1}{V_2} = \frac{R_1}{R_2},$$

i.e. R_1 and R_2 divide the total p.d. E so that the voltage ratio equals the resistance ratio.

Similarly, for the other potential divider ADC,

$$\frac{V_4}{V_3} = \frac{R_4}{R_3}.$$

Now zero current through G shows that the two potential dividers give no p.d. between B and D, i.e. they divide the total E in the same ratio,

$$\frac{V_1}{V_2} = \frac{V_4}{V_3}. \quad \therefore \frac{R_1}{R_2} = \frac{R_4}{R_3}.$$

Solution

Using the data,

$$R_1 = R_2 \times \frac{R_4}{R_3} = 200 \times \frac{1000}{100} = 2000\ \Omega.$$

Now the current through R_4 and R_3 is given by

$$I_4 = \frac{E}{R_4 + R_3} = \frac{2}{1100}\ \text{A}.$$

$$\therefore\ V_3 = I_4 \times R_3 = \frac{2}{1100} \times 100 = \frac{2}{11}\ \text{V}.$$

3.21. A Wheatstone bridge circuit is connected as in Fig. 3.8. The supply battery E has an e.m.f. of 2.0 V and a series resistance of 2.5 Ω. Calculate the resistance R_3 to balance the bridge when $R_1 = 10\ \Omega$, $R_2 = 20\ \Omega$, and $R_4 = 30\ \Omega$. Calculate also the values of V_1, V_2, V_3, and V_4 at balance. [60 Ω; 0.6 V; 1.2 V; 0.6 V; 1.2 V.]

3.22. A resistor of nominal value 30 Ω is compared with a standard resistor of 15.0 Ω by means of a metre bridge. The bridge balances when the jockey is 33.5 cm from one end of the wire. Calculate the true resistance of the resistor. [29.8 Ω.]

3.23. Calculate the resistance X and the current it carries in the circuit of Fig. 3.9, when the meter indicates zero. [108 Ω; 50 mA.]

3.24. A bridge circuit has three resistors and a voltmeter connected as a square $ABCD$: $AB = 1000\ \Omega$; $BC = 1000\ \Omega$; $CD = 1500\ \Omega$; AD is the voltmeter. A battery is connected across the diagonal AC and a tapping key across the diagonal BD. Calculate from first principles the resistance of the voltmeter if its deflection remains unchanged when the tapping key is opened and closed. [1500 Ω.]

3.25. A Wheatstone bridge circuit has four resistors P, Q, R, and S connected in series cyclic order around a square. When $P = 100\ \Omega$ and $Q = 10\ \Omega$, the bridge is balanced by shunting R with a 10Ω resistor. When P is changed to 150 Ω, the shunt must be

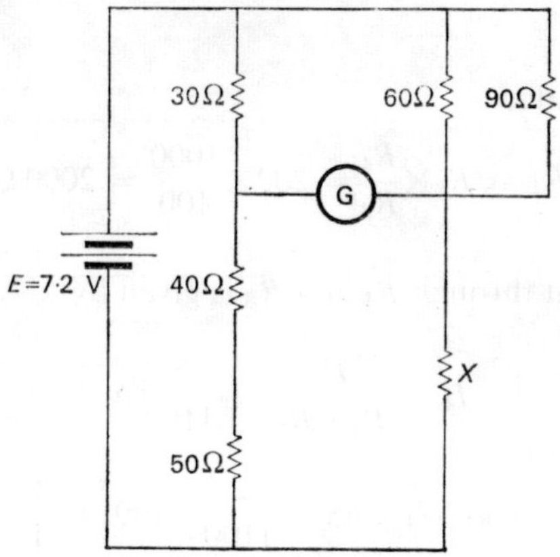

FIG. 3.9. Ex. 3.23.

changed to 5 Ω for balance. Calculate the resistances of S and R. [S = 50 Ω; R = 10 Ω.]

3.26. Four components are wired in a square *ABCD*. *AB* is a battery; *BC* is a 50 Ω resistor; *CD* is a galvanometer; *DA* is a 30 Ω resistor. A 40 Ω resistor and an unknown resistor are connected across diagonals *BD* and *AC* respectively. Calculate the unknown resistance if the galvanometer indicates zero current. [37.5 Ω.]

Distribution Circuits

3.27. WORKED EXAMPLE

A heating element of resistance 5.0 Ω is connected across a battery of e.m.f. 12 V and internal resistance 0.50 Ω. In an attempt to increase the current through the heater, a second battery, identical with the first, is connected in parallel by two leads each of resistance 0.25 Ω. Calculate the current drawn from each battery.

Introduction

If the two battery currents are a and b respectively, in the directions shown in Fig. 3.10, the current through the heater element is $a+b$. This is in accordance with Kirchhoff's first law:

That the algebraic total of current at any junction is zero when currents arriving and currents leaving are counted as positive and negative respectively.

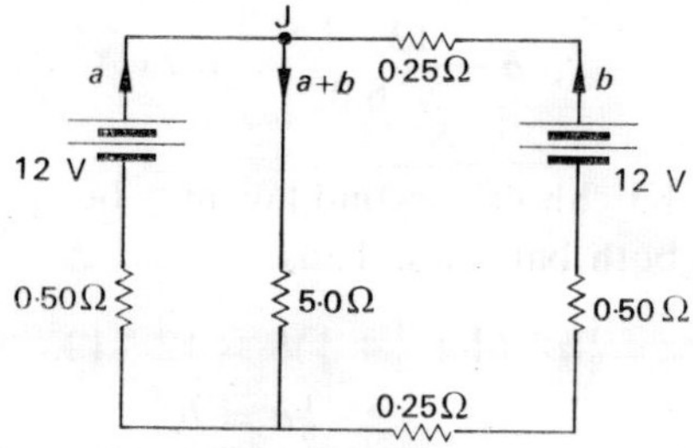

FIG. 3.10. EX. 3.27.

Thus $a+b-(a+b) = 0$ at the junction J. Note that if the value of a or b is found to be negative, it shows that the current flows in a direction opposite to that chosen.

The equations from which the values of a and b are calculated come from the application of Kirchhoff's second law:

That the resultant potential drop due to resistance around any closed path equals the resultant e.m.f. around that path.

Here the potential drops are expressed as the products of current and resistance and counted as positive or negative according to whether the current and resultant e.m.f. are in the same or opposite directions respectively.

Solution

Around the left-hand mesh of the circuit, the second law gives

$$0.50a+5.0(a+b) = 12. \quad \therefore \; 5.5a+5.0b = 12. \qquad (1)$$

Similarly, for the right-hand mesh,

$$(0.50+0.25+0.25)b+5.0(a+b) = 12. \quad \therefore \; 5.0a+6.0b = 12. \quad (2)$$

To eliminate b,

$$(1) \times 6: 33a+30b = 72. \quad (3)$$

$$(2) \times 5: 25a+30b = 60. \quad (4)$$

$$(3)-(4): 8a = 12. \quad \therefore \; a = 1.5 \text{ A}.$$

Then substituting in (2): $7.5+6.0b = 12.$

$$\therefore \; b = \frac{12-7.5}{6.0} = 0.75 \text{ A}.$$

As a check, Kirchhoff's second law may be applied to the outer path including both batteries. Thus

$$\tfrac{1}{2}a-(\tfrac{1}{4}+\tfrac{1}{2}+\tfrac{1}{4})b = 12-12.$$

$$\therefore \; \tfrac{1}{2}a = b$$

in accordance with the values calculated above.

3.28. Six lamps, each requiring a current of 2.0 A at a p.d. of 24 V, are to be run in parallel at one end of a pair of leads of resistance 0.050 Ω each, fed by a generator of internal resistance 1.0 Ω. Calculate the required e.m.f. and terminal p.d. of the generator. [37 V; 25 V.]

3.29. A battery of e.m.f. 3 V and resistance $\frac{1}{2}$ Ω is connected across a resistor of 3 Ω. Calculate the change of current in the resistor when a second battery of e.m.f. 9 V and resistance $1\frac{1}{2}$ Ω is connected in parallel with the first and with like terminals connected. What would have been the change in current if the first battery had had zero internal resistance? [+0.5 A; 0.]

3.30. The resistors R_1, R_2, and R_3 in the circuit of Fig. 3.11 are such that they draw currents equal to that in the 12 Ω resistor when a p.d. is applied between A and B. Calculate the currents drawn by these resistors when a p.d. of 24 V is applied between C and D. [0.65 A; 0.76 A; 0.92 A.]

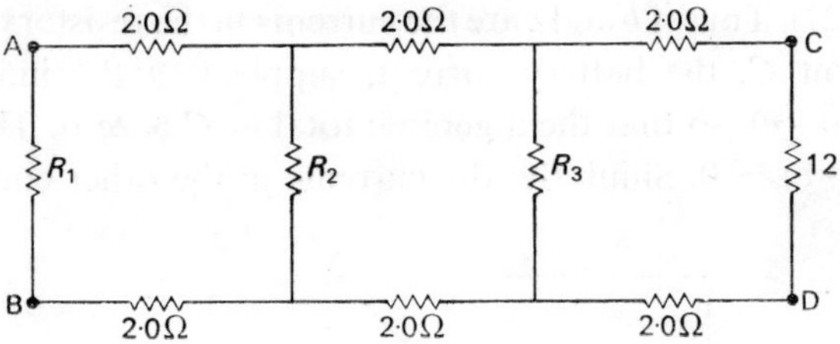

FIG. 3.11. EX. 3.30.

3.31. In a laboratory ring main circuit a twin cable of length 40 m is connected at both ends to the supply at a p.d. of 6.0 V. Three loads of resistance 1.0, 2.0, 3.0 Ω respectively are connected in this order at intervals of 10 m from the supply point. If each lead has resistance of 5.0 Ω per 1000 m, calculate the p.d. at the 2 Ω load. [5.4 V.]

Unbalanced Potentiometers and Bridges

3.32. WORKED EXAMPLE

An unbalanced Wheastone bridge circuit has three resistors of 100 Ω each and one of 99 Ω. The galvanometer resistance is 10 Ω and the supply battery e.m.f. is 2.0 V with negligible internal resistance. Calculate the galvanometer current.

Introduction

The analysis of ex. 3.20 shows that a bridge circuit is balanced only when its two potential dividers divide the applied p.d. in the same ratio. Thus in the present case the 99 Ω resistor has 1 Ω less than the resistance required for balance; so a current a flows through the galvanometer (Fig. 3.12).

To calculate a, the currents in all parts of the circuit are denoted by the minimum number of symbols using Kirchhoff's first law (see ex. 3.27). Thus if b and c are the currents in the resistors connected to point C, the battery current, supplied to the junction C, must be $(b+c)$, so that the algebraic total at C is zero. Thus at C, $b+c-(b+c) = 0$. Similarly, the currents in the other components

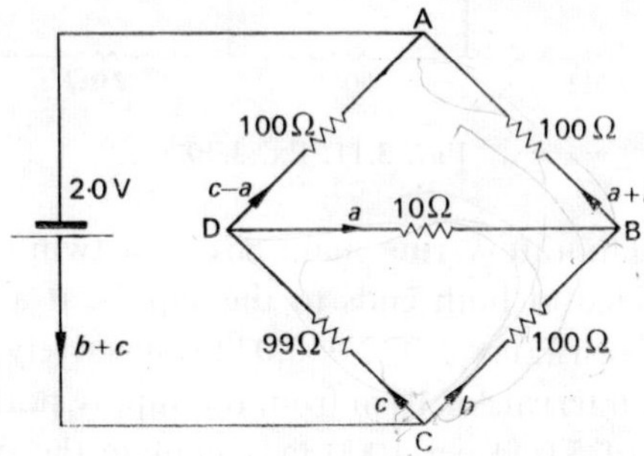

FIG. 3.12. Unbalanced Wheatstone bridge (ex. 3.32).

may be expressed in terms of a, b, and c by applying the law again at the junctions B and D. Thus at B, $a+b-(a+b) = 0$, at D, $c-a-(c-a) = 0$, and at A, $(a+b)+(c-a)-(b+c) = 0$. Note that the required current has been denoted by a single symbol a with a view to easing the calculation.

The equations, from which a, b, and c can be calculated, come from the application of Kirchhoff's second law (see ex. 3.27).

Solution

Around the outer path $CBAC$,

$$100b+100(a+b) = 2.0 \tag{1}$$

assuming no resistance in the leads carrying the current $(b+c)$.

Similarly for the path $CBDC$,

$$100b-10a-99c = 0 \tag{2}$$

and for the path $BADB$,

$$100(a+b)-100(c-a)+10a = 0. \tag{3}$$

Now from (1), $100a+200b = 2.0,$ (4)

from (2), $10a-100b+99c = 0,$ (5)

and from (3), $210a+100b-100c = 0.$ (6)

To eliminate c,

(2)×100: $1000a-10000b+9900c = 0.$ (7)

(3)×99: $20790a+9900b-9900c = 0.$ (8)

Then (7)+(8): $21790a-100b = 0.$ (9)

(9)×2: $43580a-200b = 0$

Adding (4): $100a+200b = 2.0$

$43680a = 2.0$

$$a = \frac{2.0}{43680} = 45.7\times10^{-6}\text{ A} = 46\ \mu\text{A}.$$

3.33. Calculate the current in each resistor of Fig. 3.13a. (The solution is given in Fig. 3.13b.)

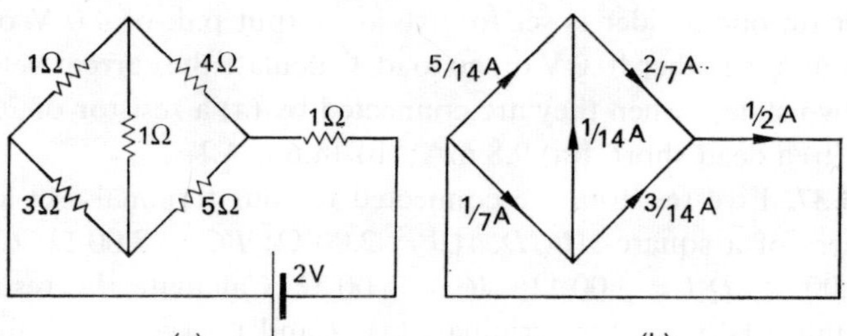

FIG. 3.13 (a) and (b). Ex. 3.33.

3.34. The e.m.f. of a cell having internal resistance of 1.0 Ω is measured on a slide-wire potentiometer which has a wire of length 100.0 cm and resistance of 10.0 Ω. The supply battery has an e.m.f. of 2.00 V and negligible internal resistance. The galvanometer has a resistance of 5.0 Ω. Balance is obtained across 75.0 cm of the wire. Calculate the galvanometer current when the jockey is set at (a) 74.0 cm, and (b) 76.0 cm. [(a) 2.53 mA; (b) 2.56 mA.]

3.35. Calculate (a) the p.d. between the points *A* and *B* in the circuit of Fig. 3.14, and (b) the current drawn by a link of negligible resistance between *A* and *B*. [(a) 0.8 V; (b) 0.1 A.]

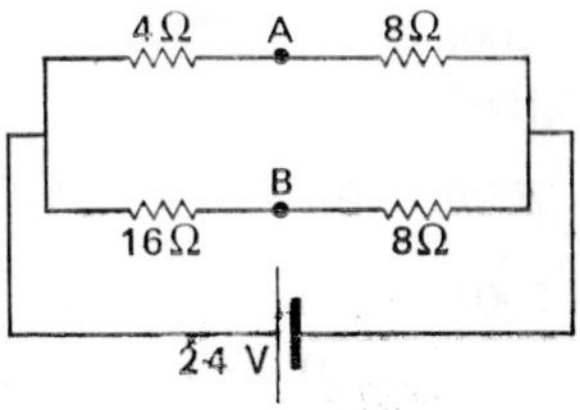

FIG. 3.14. EX. 3.35.

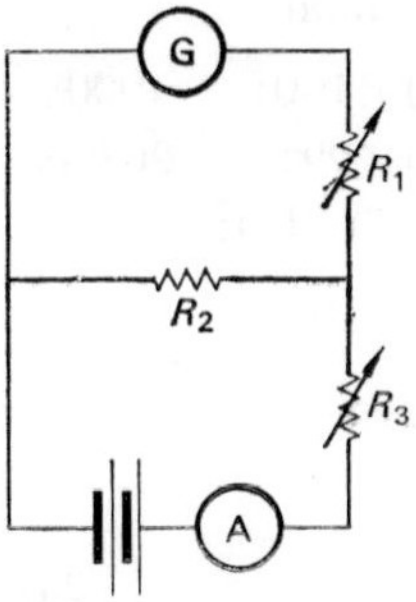

FIG. 3.15. EX. 3.38.

3.36. Two potential dividers have resistances of 1000 Ω each and are connected in parallel across a supply of p.d. 20.0 V. The slider on one divider is set to give an output p.d. of 4.0 V on no load, and the other 10.0 V on no load. Calculate the current between the two sliders when they are connected by (a) a resistor of 200 Ω, and (b) a dead short. [(a) 9.8 mA; (b) 14.6 mA.]

3.37. Five resistors are connected to four terminals set at the corners of a square *ABCD*. *AB* = 2.00 Ω; *BC* = 3.00 Ω; *CD* = = 2.00 Ω; *DA* = 3.00 Ω; *AC* = 5.00 Ω. Calculate the resultant resistance between the terminals (a) *A* and *C*, and (b) *B* and *D*. [(a) 1.67 Ω; (b) 2.47 Ω.]

3.38. The circuit shown in Fig. 3.15 is used in the calibration of a galvanometer *G*. When $R_1 = 0$, $R_2 = 0.10\,\Omega$, and $R_3 = 1.5\,\text{k}\Omega$, the milliammeter *A* indicates 24 mA and *G* has a deflection of 72 mm. If R_1 is then increased from zero to 15 Ω, the deflection of *G* becomes 36 mm. Calculate for the galvanometer (a) the voltage sensitivity, and (b) the current sensitivity. Use approximations consistent with an accuracy of two significant figures. [(a) 30 mm mV^{-1}; (b) 450 mm mA^{-1}.]

CHAPTER 4

POWER

Power

4.1. Worked Example

Two resistors, of 50.0 kΩ and 150 kΩ respectively, are connected (a) in parallel, and (b) in series with a supply of 200 V e.m.f. In each case calculate the power drawn from the source and the power dissipated by each resistor.

Introduction

It was shown in ex. 1.10 that when a charge Q is driven by an e.m.f. E around a circuit, the energy given by the source of e.m.f. is $W = EQ$. Similarly, the energy leaving the circuit between points at a p.d. V is VQ.

Now, in general, power is the rate of change of energy. In this case the change is from the energy of the source of e.m.f. to heat energy dissipated by the resistors.

So when energy W is supplied by the source of e.m.f. in time t, the input power to the circuit is

$$P = \frac{W}{t} = \frac{EQ}{t},$$

where E is in volts, Q in coulombs, t in seconds, W in joules, and P in watts. But Q/t = the current I in amperes.

$$\therefore\ P = EI. \tag{4.1}$$

Similarly, the output power between points at a p.d. V is VI. So for a p.d. V across a resistor R carrying a current $I = V/R$,

$$\text{output power } P = VI = V^2/R = I^2R, \tag{4.2}$$

where V is in volts, I in amperes, R in ohms, and P in watts.

Solution

In part (a) of the question, each parallel resistor R_1 and R_2 (Fig. 4.1) has a p.d. V equal to the e.m.f. E.

Then eqn. (4.2) gives:

$$P_1 = \frac{V^2}{R_1} = \frac{200^2}{50{,}000} = 0.80 \text{ W}$$

and

$$P_2 = \frac{V^2}{R_2} = \frac{200^2}{150{,}000} = 0.27 \text{ W}.$$

Therefore the total power drawn from the source and dissipated by the resistors is $P_1+P_2 = 1.07$ W.

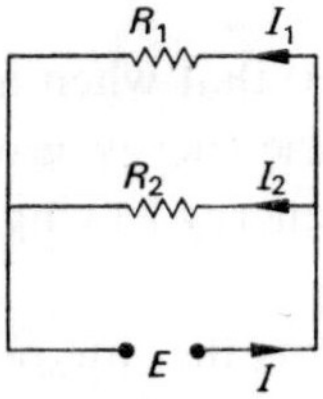

FIG. 4.1. Ex. 4.1(a).

Alternatively, the total power may be calculated from the total current

$$I = I_1+I_2 = \frac{E}{R_1}+\frac{E}{R_2} = \frac{200}{50{,}000}+\frac{200}{150{,}000}$$

$$= 5.33\times10^{-3} \text{ A}.$$

Then, from eqn. (4.1),

$$P = EI = 200\times5.33\times10^{-3} = 1.07 \text{ W}.$$

(b) Secondly, in Fig. 4.2 the total resistance in series is $R = R_1 + R_2 = 50 + 150\ \text{k}\Omega = 2 \times 10^5\ \Omega$.

$$\therefore \text{ current } I = \frac{E}{R} = \frac{200}{2 \times 10^5} = 10^{-3}\ \text{A}.$$

Then, from eqn. (4.1), power drawn from source

$$P = EI = 200 \times 10^{-3} = 0.20\ \text{W}.$$

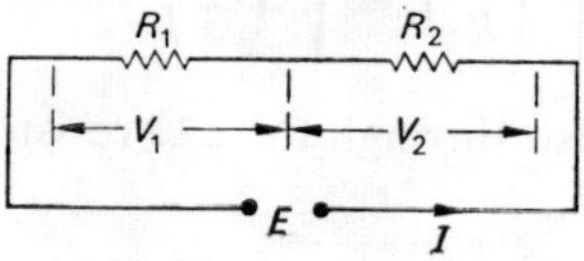

FIG. 4.2. Ex. 4.1(b).

From eqn. (4.2), the powers P_1 and P_2 dissipated by resistors R_1 and R_2 are

$$P_1 = I^2 R_1 = (10^{-3})^2 \times 50 \times 10^3 = 0.05\ \text{W}$$

and $$P_2 = I^2 R_2 = (10^{-3})^2 \times 150 \times 10^3 = 0.15\ \text{W}.$$

Note that $P = P_1 + P_2$.

4.2. WORKED EXAMPLE

In the circuit given by Fig. 4.3, a power of 10 W is dissipated by the 3 Ω resistor. Calculate (a) the power in each other resistor, and (b) the total energy in joules and kilowatt-hours taken from the source in 25 min.

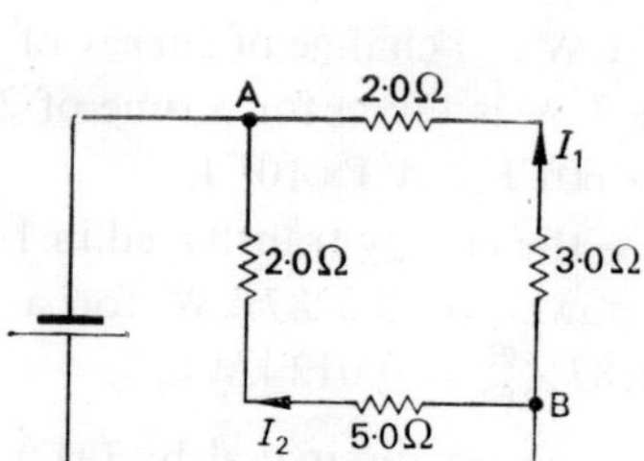

FIG. 4.3. Ex. 4.2.

Solution

From eqn. (4.2),

$$P = I^2R. \quad \therefore \; I = \sqrt{\left(\frac{P}{R}\right)}.$$

Therefore the current through the 3 Ω resistor is

$$I_1 = \sqrt{\left(\frac{P_1}{R_1}\right)} = \sqrt{\left(\frac{10}{3}\right)} = 1.83 \text{ A.}$$

This current passes also through the 2 Ω resistor in series, in which the power is

$$I_1^2 \times 2 = \tfrac{10}{3} \times 2 = 6.7 \text{ W.}$$

Then the p.d. applied between A and B across 2 Ω and 3 Ω in series is

$$I_1(2+3) = 1.83 \times 5 = 9.15 \text{ V.}$$

The same p.d. across 2 Ω and 5 Ω in series gives the current

$$I_2 = \frac{9.15}{7} = 1.31 \text{ A.}$$

Therefore the powers dissipated by the 2 Ω and 5 Ω resistors are

$$1.31^2 \times 2 = 3.4 \text{ W} \quad \text{and} \quad 1.31^2 \times 5 = 8.6 \text{ W.}$$

Then the total power drawn $= 10+6.7+3.4+8.6 = 28.7$ W.
Alternatively, the total power $= V(I_1+I_2) = 9.15 \times (1.83+1.31)$
$= 28.7$ W.

Now a power of 1 W is a change of energy of 1 J s^{-1}. Therefore when a power of 28.7 W is taken for a time of 25 min, the energy taken $= 28.7 \times (25 \times 60)$ J $= 4.3 \times 10^4$ J.

Further, 1 kWh is the energy transferred in 1 h when the power is 1 kW. So for a power of 0.0287 kW for a time of $\frac{25}{60}$ h, the energy taken $= 0.0287 \times \frac{25}{60} = 0.012$ kWh.

4.3. Calculate the power dissipated by (a) a resistor of 1.0 kΩ carrying a current of 50 mA, and (b) a resistor of 2.0 MΩ having a p.d. of 2.0 kV. Calculate also the p.d. across (c) a resistor of

25 kΩ dissipating a power of 10 W, and (d) a resistor dissipating a power of 10 mW when carrying a current of 50 μA. [(a) 2.5 W; (b) 2 W; (c) 500 V; (d) 200 V.]

4.4. Calculate the resistance of (a) an electric heater rated at 5.0 kW, 240 V, and (b) supply leads which dissipate a power of 2.0 W when carrying a current of 13 A. Calculate also the energy dissipated in 24 h in each case, giving the answers in both joules and kilowatt-hours. [(a) 12 Ω, 4.3×10^8 J, 120 kWh; (b) 0.012 Ω, 1.7×10^5 J, 0.048 kWh.]

4.5. A heater element has a centre tap enabling the halves to be used (a) singly, (b) in series, (c) in parallel. Calculate the power drawn from the same source in cases (a) and (c) if that of (b) is 1 kW. [(a) 2 kW; (c) 4 kW.]

4.6. A 12 V, 6 W lamp is to be run from a battery of e.m.f. 15 V and internal resistance 1 Ω. Calculate the resistance and power rating of the series resistor required for correct operation. [5 Ω; $1\frac{1}{4}$ W.]

4.7. A 75 W lamp, a 500 W laundry iron, and a 5.0 kW water heater run continually in parallel for 2 h on a 240 V supply at a cost of 1.8p per kilowatt-hour. Calculate (a) the total current, (b) the charge passed, (c) the energy taken in joules, and (d) the total cost. [(a) 23 A; (b) 1.7×10^5 C; (c) 4×10^7 J; (d) 20p.]

4.8. Three resistors of resistance 60 Ω, 100 Ω, and 150 Ω respectively are connected in parallel. A supply of 24 V is to be used to provide a power of $2\frac{1}{4}$ W in the resistor of 100 Ω. (a) Calculate the resistance of the resistor to be connected in series with the supply and the parallel group, and (b) determine which resistor dissipates most power in this arrangement. [(a) 18 Ω; (b) 18 Ω.]

4.9. Two resistors in series are to be used as a potential divider across a battery of e.m.f. 150 V and negligible internal resistance, to supply a current of 30 mA at a p.d. of 100 V. (a) Calculate the values of the resistances required if the smaller resistor is to dissipate a power of 2.0 W. (b) Calculate also the power dissipated by the other resistor. [(a) 1.25 kΩ, 10 kΩ; (b) 1 W.]

4.10. A heater of power 10 kW is connected to a supply by leads of total resistance 0.10 Ω. Calculate the power loss in the leads when the heater operates correctly at a p.d. of (a) 200 V, and (b) 500 V. Calculate also (c) the minimum p.d. at which the heater must operate if the power loss in the leads is not to exceed 1% of the heater power. [(a) 250 W; (b) 40 W; (c) 316 V.]

4.11. Apparatus requiring a power supply of 64 W is to be run from a 24 V supply with a suitable series resistor of resistance R. When $R = 4\ \Omega$, the power supplied is 36 W. Calculate the correct value of R. [2 Ω.]

4.12. A battery has e.m.f. of 4.50 V and internal resistance of 1.50 Ω. Calculate the power drawn by a lamp of resistance (a) 1.00 Ω, and (b) 3.00 Ω. Also determine (c) the maximum power which can be drawn by a lamp from this battery. [(a) 3.24 W; (b) 3.00 W; (c) 3.38 W.]

Heating

4.13. Worked Example

A 2 kW electric kettle holds a mass of 2.0 kg of water which is boiled from an initial temperature of 15°C with an efficiency of 75%. Calculate (a) the time taken, and (b) the heat lost to the surroundings. Ref.: water.

Solution

The heat in joules required to raise the temperature of a body $= mc\ \Delta\theta$, where m is the mass in kg, c is the specific heat in J kg^{-1} deg C^{-1}, and $\Delta\theta$ is the rise of temperature in deg C. Here $m = 2.0$ kg, $\Delta\theta = 100-15 = 85$ deg C, and c is given in the reference section as 4.19×10^3 J kg^{-1} deg C^{-1}.

$\therefore$ heat gained by water $= 2.0\times4.19\times10^3\times85 = 7.12\times10^5$ J.

This quantity is 75% of the total heat H supplied by the heater element, the other 25% being lost.

$$\therefore\ 7.12\times10^5 = 0.75\times H. \quad \therefore\ H = 9.5\times10^5\ \text{J}.$$

Now 2 kW is a power of 2000 W = 2000 J s^{-1}. Therefore the required time

$$t = H/2000 = 9.5\times10^5/2000 = 475\ \text{s} = 7.9\ \text{min}. \quad \text{(a)}$$

$$\text{The heat lost} = 25\%\ \text{of}\ H = 0.25\times9.5\times10^5 = 2.4\times10^5\ \text{J}. \quad \text{(b)}$$

4.14. Calculate the rise of temperature of a mass of 500 g of water when it is stirred for 20 min by a paddle driven by an electric motor taking a current of 2.0 A at a p.d. of 12 V. Assume an efficiency of $33\frac{1}{3}\%$. Ref.: water. [4.6 deg C.]

4.15. Calculate the resistance of a 240 V electric kettle which brings 1.0 l of water to the boil from a temperature of 10°C in a time of 10 min with efficiency of 75%. Ref.: water. [69 Ω.]

4.16. A well-lagged cylindrical tank of length 1.0 m and diameter 0.30 m is full of water and is fitted with an immersion heater rated at 3.0 kW, 240 V. Calculate (a) the resistance of the heater element, (b) the time required for a temperature rise of 10 deg C, and (c) the temperature rise obtained in 10 min. Ref.: water. [(a) 19 Ω; 16 min.; 6.1 deg C.]

4.17. A mass of 50 kg of iron rests in an electric oven which takes a current of 16 A from a 250 V supply. The thermal capacity of the oven itself and the heat losses from the oven take 25% of the heat produced in 30 min. Calculate the energy taken from the supply and the rise of temperature of the iron in 30 min. Assume that the specific heat of iron is 470 J kg^{-1} deg C^{-1}. [2.0 kWh; 230 deg C.]

4.18. Calculate the voltage drop and the heat loss in joules per minute in leads of total series resistance 0.023 Ω, supplying a power of 6.0 kW at a p.d. of 250 V measured at the load. [0.55 V; 800 J/min.]

4.19. A 240 V immersion heater in 30 kg of water produces a rise of temperature of 19.1 deg C in 20 min. Calculate (a) the power taken, (b) the charge passed, and (c) the cost at a rate of 2p per kWh. Assume 100% efficiency. Ref.: water. [(a) 2 kW; (b) 10^4 C; (c) 1.3p.]

4.20. A copper can of mass 250 g contains 1.0 kg of paraffin oil. It is heated and allowed to cool. At 50°C the rate of cooling is 1.5 deg C min^{-1}. Calculate the current in a fully immersed heater of resistance 100 Ω to maintain the temperature at 50°C. Assume the specific heat of paraffin oil to be 2200 J kg^{-1} deg C^{-1}. Ref.: copper. [0.76 A.]

4.21. An electric hot plate rated at 5.0 A, 250 V heats a kettle with efficiency of 65%. Calculate the time taken to boil dry starting with $1\frac{1}{2}$ kg of water at 20°C. Assume that the specific heat of water is 4190 J kg^{-1} deg C^{-1} and the latent heat of steam is 2.25×10^6 J kg^{-1}. [80 min.]

Decibels

4.22. Worked Example

Calculate the power ratios given by (a) 3 dB, and (b) −2 dB. Also express in decibels the power ratios (c) 1 : 5, (d) 1 : 10^6, and (e) 1 : 5×10^6.

Introduction

If a ratio of two powers P_1 and P_2 is expressed in index form to a base of 10, then the index is a measure of the ratio in "bels". Thus if $P_2/P_1 = 10^n$, then P_2 is greater than P_1 by n bels.

Invariably this unit is reduced by a factor of $\frac{1}{10}$ to give a decibel (dB). Thus n bels = $10n$ dB = N dB.

So, for a power ratio of N dB, $P_2/P_1 = 10^{N/10}$, and $N = 10 \log_{10} (P_2/P_1)$ dB.

It follows that 1 dB is a power ratio of $\sqrt[10]{10} = 1.26$, and N dB is a power ratio of $(1.26)^N$.

Solution

(a) 3 dB is the ratio $P_2/P_1 = 10^{0.3}$.

But from antilog tables, $10^{0.3} = 1.995 = 2.0$ (2 sig. fig.).

$\therefore$ 3 dB is a ratio of 1 : 2 approximately. (a)

(b) -2 dB is the ratio $P_2/P_1 = 10^{-0.2} = 10^{\bar{1}.8} = 0.63$.

$\therefore$ -2 dB is a ratio of 1 : 0.63. (b)

(c) When $P_2/P_1 = 5$, $N = 10 \log_{10} 5 = 6.99$ dB

$= 7.0$ dB (2 sig. fig.). (c)

(d) When $P_2/P_1 = 10^6$, $N = 60$ dB. (d)

(e) When $P_2/P_1 = 5\times10^6$,

$$
\begin{aligned}
N &= 10 \log_{10}(5\times10^6)\\
&= 10(\log_{10} 5+\log_{10} 10^6)\\
&= 10(0.699+6)\\
&= 67 \text{ dB} \quad \text{(2 sig. fig.).}
\end{aligned}
$$
(e)

Alternatively, (e) may be obtained from (c) and (d) by using a common property of indices: $10^{0.7}\times10^6 = 10^{6.7}$

or using dB: $1.26^7\times1.26^{60} = 1.26^{67}$.

4.23. A current of 100 mA flows through a resistor of 10 kΩ in series with a resistor of 100 Ω. Calculate the ratio of the two powers *A* and *B* respectively in decibels. [$A/B = 20$ dB.]

4.24. Calculate in decibels the ratio of powers developed in the following cases of currents in resistors: (a) 2 mA, 50 kΩ; (b) 5 mA, 20 kΩ. [(a)/(b) = -4 dB.]

4.25. A p.d. of 5.0 V across a resistor of 1.0 MΩ (1) is to be compared with a p.d. of 10 V across a resistor of 10 kΩ (2). Calculate the power ratio in decibels. [(1)/(2) = -26 dB.]

4.26. Calculate the powers, which, relative to a power of 1 mW, are given by the following series: -4.0 dB, -2.0 dB, 0 dB, $+2.0$ dB, $+4.0$ dB. [400 μW; 630 μW; 1.0 mW; 1.6 mW; 2.5 mW.]

4.27. Convert the following series of powers from watts to decibels relative to a power of 100 W: 50 W, 100 W, 150 W, 200 W. [−3.0 dB; 0 dB; +1.8 dB; +3.0 dB.]

4.28. At various times the power developed in a resistor is given by the following series: (a) 3.0 W, 9.0 W, 27 W, 81 W, and (b) 3.0 W, 6.0 W, 9.0 W, 12 W. Convert each series from watts to decibels relative to the initial value. [(a) 0 dB, 4.8 dB, 9.5 dB, 14 dB; (b) 0 dB, 3.0 dB, 4.8 dB, 6.0 dB.]

4.29. A resistor of 10 Ω draws a current of 300 mA. Calculate the p.d. across the resistor when the power drawn is increased by 5.0 dB. [5.3 V.]

CHAPTER 5

RESISTIVITY AND TEMPERATURE

Resistivity

5.1. Worked Example

A flexible copper lead has 23 parallel wires of circular cross-section each of diameter 0.019 cm and length 3.0 m. The resistivity of the copper $= 1.7 \times 10^{-8}\,\Omega\text{m}$. Calculate the resistance of the lead.

Introduction

The resultant resistance of n conductors connected in parallel is given by adding n reciprocals, as in eqn. (2.8) where n was 3. Thus

$$1/R_n = n/R, \tag{5.1}$$

where R_n is the resultant resistance of n parallel conductors or resistors each of resistance R.

An increase of cross-section area A of a single conductor is equivalent to the connection of additional conductors in parallel. So R depends on A just as R_n depends on n. But $1/R_n \propto n$ in eqn. (5.1).

$$\therefore\ 1/R \propto A \quad \text{and} \quad R \propto 1/A. \tag{5.2}$$

Furthermore, an increase of length l of a single conductor is equivalent to the connection of additional conductors in series, in which case the resistances add directly.

So

$$R \propto l, \tag{5.3}$$

just as $R_n \propto n$ for n equal resistors in series.

Now combining the relationships (5.2) and (5.3), $R \propto l/A$, or

$$R = \varrho l/A, \tag{5.4}$$

where ϱ is a constant of proportionality called the resistivity of the material for the particular temperature.

Rearranging eqn. (5.4) gives $\varrho = RA/l$, which shows that the units of ϱ in the M.K.S. system must be given by

$$\text{ohm (metre)}^2/\text{metre} = \text{ohm metre } (\Omega\text{m}).$$

Solution

In the particular problem, the

resistivity $\varrho = 1.7\times10^{-8}$ Ωm,
diameter $d = 1.9\times10^{-4}$ m,
area $A = \pi d^2/4 = \pi(1.9\times10^{-4})^2/4 = 2.84\times10^{-8}$ m^2,
length $l = 3.0$ m.

For each wire, eqn. (5.4) gives

$$R = \varrho l/A = 1.7\times10^{-8}\times3.0/(2.84\times10^{-8}) = 1.8\ \Omega.$$

Then for 23 wires in parallel, eqn. (5.1) gives

$$R_n = R/n = 1.8/23 = 7.8\times10^{-2}\ \Omega.$$

5.2. Calculate the length of wire of size No. 32 S.W.G. to give a resistance of 5.0 Ω, assuming a resistivity of 44×10^{-8} Ωm. No. 32 S.W.G. is a standard wire gauge of diameter 0.274 mm. [67 cm.]

5.3. A rectangular block of metal measures 6.0×2.0×3.0 cm. Calculate the resistance (a) between opposite ends, and (b) between opposite sides. Assume a resistivity of 1.8×10^{-8} Ωm. [(a) 1.8 μΩ; (b) 0.20 μΩ or 0.45 μΩ.]

5.4. A coil has a mean diameter of 5.0 cm and is wound with 250 turns of copper wire of diameter 0.50 mm. Calculate the resistance. Ref.: copper. [3.4 Ω.]

5.5. Calculate the resistance of an aluminium tube having a length of 15 m, an internal diameter of 2.5 cm, and an external diameter of 3.0 cm. Ref.: aluminium. [0.0019 Ω.]

5.6. The resistance of a wire of a particular material and of diameter 0.093 cm is stated to be 25.8 Ω per 1000 m. Calculate the resistance of 1.5 m of wire of the same material and of diameter 0.031 cm. [0.35 Ω.]

Mass of Conductor

5.7. Worked Example

Two uniform conductors have equal lengths and resistances but are of different materials—aluminium and copper. Calculate the ratio of the masses of the two conductors. Ref.: copper and aluminium.

Introduction

As shown in eqn. (5.4), resistance $R = \varrho l/A$, where ϱ = resistivity, l = length, and A = cross-section area of a conductor or resistor. Also, for the same piece of material the mass $m = dAl$, where d = density. Note that confusion would arise if the symbol ϱ were used for both resistivity and density.

Now for equal resistances R of the two conductors (1) and (2), with equal lengths l,

$$R = \varrho_1 l/A_1 = \varrho_2 l/A_2.$$

$$\therefore\ A_1/A_2 = \varrho_1/\varrho_2.$$

Then the required ratio of masses

$$m_1/m_2 = (d_1 A_1 l)/(d_2 A_2 l) = (d_1 \varrho_1)/(d_2 \varrho_2).$$

Solution

From the reference table:

for aluminium $d_1 = 2.7\times10^3$ kg m^{-3}, $\varrho_1 = 2.8\times10^{-8}$ Ωm;

for copper $d_2 = 8.9\times10^3$ kg m^{-3}, $\varrho_2 = 1.7\times10^{-8}$ Ωm.

Therefore mass of aluminium/mass of copper

$$= (2.7\times2.8)/(8.9\times1.7) = 0.50.$$

5.8. Calculate the resistance of 1.0 kg of copper wire of diameter (a) 1.0 mm, and (b) 2.0 mm. Ref.: copper. [(a) 3.1 Ω; (b) 0.19 Ω.]

5.9. Calculate the resistance at a temperature of 20°C for 1 km of copper wire weighing 5 kg per km. Ref.: copper. [30 Ω.]

5.10. A length of aluminium wire has a diameter of 1.0 mm, a mass of 100 g, and a resistance of 1.7 Ω. Calculate (a) the resistivity, and (b) the resistance of a length of wire of the same material and mass having a diameter of 0.10 mm. Ref.: aluminium. [(a) 2.8×10^{-8} Ωm; (b) 17 kΩ.]

Resistivity, Current, and Power

5.11. Worked Example

The tungsten filament of an experimental lamp rated at 150 W, 240 V, has a diameter of 0.050 mm. Calculate the power dissipated per unit area of the filament surface, assuming a resistivity of 8.0×10^{-7} Ωm at the operating temperature.

Solution

The lamp filament dissipates all power P supplied to it when its temperature is constant. The surface area of the filament can be calculated if the length is found first by considering the resistance $R = V^2/P$, where $P = 150$ W, $V = 240$ V, and $R = \varrho l/A$, where $\varrho = 8.0 \times 10^{-7}$ Ωm and $A = \pi(0.025 \times 10^{-3})^2 = 1.96 \times 10^{-9}$ m².

$$\therefore l = AV^2/\varrho P = 1.96 \times 10^{-9} \times 240^2/(8.0 \times 10^{-7} \times 150)$$
$$= 0.944 \text{ m}.$$

Then

$$\text{surface area} = \text{length} \times \text{perimeter of cross-section}$$
$$= l\pi d = 0.944 \times \pi \times 0.050 \times 10^{-3} = 0.148 \times 10^{-3} \text{ m}^2.$$

Therefore power dissipated per unit area $= 150/(0.148 \times 10^{-3})$

$$= 1.0 \times 10^6 \text{ W m}^{-2}$$
$$= 1.0 \text{ W mm}^{-2}.$$

5.12. A heater element rated at 240 V, 750 W is to be made of wire of diameter 0.56 mm and resistivity 1.2 μΩm. Calculate the length of wire required. [16 m.]

5.13. Fuse wires of a certain material are tested under the same conditions of cooling. Wires of diameters 0.152 mm and 0.376 mm are found to fuse for currents of 5.0 A and 20 A respectively. Calculate the fusing current for a wire of 0.315 mm diameter. Assume that the fusing current is proportional to a constant power of the diameter. [15 A.]

5.14. A number of lamps taking a total current of 20 A are supplied from 240 V mains by twin copper leads of diameter 0.20 cm. Calculate the maximum length of twin lead if the total volt drop at 20°C is not to exceed 2% of the supply p.d. Ref.: copper. [22 m.]

5.15. Calculate the weight of copper conductors needed for a direct supply to a load of 10 kW at a distance of 1.0 km. The power loss in the leads at 20°C is not to exceed 4% of the load power when the load p.d. is (a) 250 V, and (b) 1000 V. Ref.: copper. [(a) 2400 kg; (b) 150 kg.]

5.16. The wire of a heater element of resistance 40 Ω, for use on a 240 V supply, is to be reduced in length so as to run on a 200 V supply. Calculate the proportion of length which must be removed if the lower voltage is to give (a) the same current, (b) the same power, and (c) the same power per unit area of surface. [(a) 17%; (b) 30%; (c) 17%.]

Temperature Coefficient at 0°C

5.17. Worked Example

A coil is wound with 30 m of copper wire of cross-section area 0.10 mm². Calculate the resistance at a temperature of (a) 0°C, and (b) 50°C. The resistivity of copper at 20°C is 1.7×10^{-8} Ωm and the temperature coefficient of resistance at 0°C is 4.3×10^{-3} deg C^{-1}.

Introduction

When a metal is heated, its resistance increases approximately in proportion to the temperature rise. Thus, assuming exact proportion

$$\Delta R = \alpha_0 R_0 \, \Delta\theta, \tag{5.5}$$

where ΔR = increase of resistance, R_0 = resistance at 0°C, $\Delta\theta$ = rise of temperature, and α_0 = temperature coefficient of resistance at 0°C.

Now if R_θ is the resistance at θ°C, $\Delta R = R_\theta - R_0$ and $\Delta\theta = \theta - 0$.

Then, from eqn. (5.5),

$$R_\theta - R_0 = \alpha_0 R_0 \theta \quad \text{or} \quad R_\theta = R_0(1+\alpha_0\theta). \tag{5.6}$$

This change of resistance with temperature is due primarily to a change of resistivity, the expansion of length and cross-section being negligible. So $R \propto \varrho$ where ϱ = resistivity.

Then, using eqn. (5.6),

$$\varrho_\theta = \varrho_0(1+\alpha_0\theta). \quad (5.7)$$

Solution

(a) The resistance at 20°C is given by substituting in eqn. (5.4),

i.e. $$R_{20} = \varrho_{20}\frac{l}{A} = \frac{1.7\times10^{-8}\times30}{0.1\times10^{-6}} = 5.1\,\Omega.$$

Then, using eqn. (5.6),

$$R_{20} = R_0[1+(\alpha_0\times20)].$$
$$\therefore\ 5.1 = R_0[1+(4.3\times10^{-3}\times20)].$$
$$\therefore\ R_0 = \frac{5.1}{1+0.086} = 4.7\,\Omega. \quad \text{(a)}$$

(b) Using eqn. (5.6) again, with the value of R_0 just obtained,

$$R_{50} = R_0[1+(\alpha_0\times50)]$$
$$= 4.7[1+(4.3\times10^{-3}\times50)]$$
$$= 4.7\times1.215 = 5.7\,\Omega. \quad \text{(b)}$$

Alternatively, eqn. (5.7) may be used to calculate the resistivities at the temperatures 0°C and 50°C.

Thus for (a), $$\varrho_{20} = \varrho_0[1+(\alpha_0\times20)]$$
$$\therefore\ 1.7\times10^{-8} = \varrho_0[1+(4.3\times10^{-3}\times20)].$$
$$\therefore\ \varrho_0 = \frac{1.7\times10^{-8}}{1.086} = 1.565\times10^{-8}\,\Omega\text{m}.$$

Then $$R_0 = \varrho_0\frac{l}{A} = \frac{1.565\times10^{-8}\times30}{0.1\times10^{-6}} = 4.7\,\Omega. \quad \text{(a)}$$

Similarly for (b), $$\varrho_{50} = \varrho[1+(\alpha_0\times50)]$$
$$= 1.565\times10^{-8}[1+(4.3\times10^{-3}\times50)]$$
$$= 1.565\times10^{-8}\times1.215 = 1.9\times10^{-8}\,\Omega\text{m}.$$

Then $$R_{50} = \varrho_{50}\frac{l}{A} = \frac{1.9\times10^{-8}\times30}{0.1\times10^{-6}} = 5.7\,\Omega. \quad \text{(b)}$$

5.18. A coil of copper wire has a resistance of 25 Ω at 0°C. Calculate (a) the resistance at 40°C, and (b) the temperature at which the resistance would be 24 Ω. Ref.: copper. [(a) 29 Ω; (b) −9.3°C.]

5.19. The resistance of a coil of copper wire at a temperature of 16°C is 150 Ω. Calculate the resistance at a temperature of 50°C. Ref.: copper. [170 Ω.]

5.20. A length of 5.0 m of nickel wire, of diameter 0.26 mm, has a resistance of 6.3 Ω at 0°C. Calculate the resistivity at (a) 0°C, and (b) 100°C, assuming the temperature coefficient of resistance at 0°C to be 0.0054 deg C^{-1}. [(a) 6.7×10^{-8} Ωm; (b) 10×10^{-8} Ωm.]

5.21. An aluminium conductor of diameter 1.0 mm and length 100 m carries a current of 2.5 A. Calculate the power loss from the conductor at a temperature of (a) 0°C, and (b) 50°C. Ref.: aluminium. [(a) 21 W; (b) 25 W.]

5.22. A voltmeter, having a copper coil of resistance 100 Ω at a temperature of 10°C, is to be fitted with a multiplier of 900 Ω for use at 10°C. Calculate the voltmeter error at 30°C if the multiplier is made of (a) copper, and (b) nichrome.

Assume the temperature coefficients of resistance at 0°C to be 4.3×10^{-3} deg C^{-1} and 4.0×10^{-4} deg C^{-1} for (a) and (b) respectively. [(a) −7.6%; (b) −1.5%.]

Coefficient at Any Temperature

5.23. Worked Example

The resistance of a conductor is 76.0 Ω at a temperature of 20.0°C and 130.0 Ω at 200°C. Assuming a linear variation of resistance with temperature, calculate (a) the temperature coefficient of resistance at 20.0°C, (b) the temperature coefficient of resistance at 0°C, and (c) the resistance at 160°C.

Introduction

The temperature coefficient of resistance relative to any particular temperature is defined by an equation of the same form as eqn. (5.5). Thus

$$\Delta R = \alpha_r R_r \,\Delta\theta, \tag{5.8}$$

where ΔR = increase of resistance due to rise of temperature by $\Delta\theta$, and α_r = temperature coefficient of resistance relative to the temperature at which the resistance is R_r.

The rate of change $\Delta R/\Delta\theta$ is constant for a linear variation. Therefore $\alpha_r R_r$ = a constant, and α_r depends on R_r which depends on the reference temperature.

Equation (5.5) of ex. 5.17, i.e. $\Delta R = \alpha_0 R_0 \,\Delta\theta$, is the special case of eqn. (5.8) when the reference temperature is 0°C. Similarly, eqn. (5.6) of ex. 5.17 is a special case of the general equation

$$R_\theta = R_r[1+\alpha_r(\theta-\theta_r)], \tag{5.9}$$

where R_θ = resistance at temperature θ, and R_r = resistance at reference temperature θ_r.

Now for a reference temperature of 20°C, eqn. (5.9) is

$$R_\theta = R_{20}[1+\alpha_{20}(\theta-20)]. \tag{5.10}$$

Here α_{20} is the temperature coefficient of resistance relative to 20°C.

Solution

(a) Using the data of the question, eqn. (5.10) gives

$$130.0 = 76.0[1+\alpha_{20}(200-20)].$$

$$\therefore \alpha_{20} = \frac{130.0-76.0}{76.0\times 180} = 3.95\times 10^{-3} \text{ deg C}^{-1}. \tag{a}$$

(b) The temperature coefficient relative to 0°C is α_0 in the special case of eqn. (5.9):

$$R_\theta = R_0(1+\alpha_0\theta). \tag{5.11}$$

Substituting the given values:

$$76.0 = R_0[1+(\alpha_0\times 20)] = R_0+20R_0\alpha_0, \tag{5.12}$$

$$130.0 = R_0[1+(\alpha_0\times 200)] = R_0+200R_0\alpha_0. \tag{5.13}$$

Now multiplying eqn. (5.12) by 10 and subtracting eqn. (5.13):

$$\begin{aligned} 760 &= 10R_0+200R_0\alpha_0 \\ 130 &= R_0+200R_0\alpha_0 \\ \hline 630 &= 9R_0 \\ \hline \therefore R_0 &= 70.0\ \Omega. \end{aligned}$$

Then, returning to eqn. (5.12),

$$76.0 = 70.0+(20\times 70.0\times\alpha_0).$$

$$\therefore \alpha_0 = \frac{76.0-70.0}{20\times 70.0} = \frac{6.0}{1400} = 4.3\times 10^{-3}\ \text{deg C}^{-1}. \qquad \text{(b)}$$

Alternatively, one coefficient may be obtained from the other by the relationship developed below. Assuming $\Delta R/\Delta\theta$ to be constant, eqn. (5.8) gives $\alpha_a R_a = \alpha_b R_b$ for any two reference points a and b.

But $R_b = R_a[1+\alpha_a(\theta_b-\theta_a)]$ from eqn. (5.9).

$$\therefore \frac{\alpha_a}{\alpha_b} = \frac{R_b}{R_a} = 1+\alpha_a(\theta_b-\theta_a). \qquad \therefore \frac{1}{\alpha_b} = \frac{1}{\alpha_a}+(\theta_b-\theta_a).$$

Now putting $\theta_b = 0$, $\theta_a = 20$, and $\alpha_a = 3.95\times 10^{-3}$ deg C^{-1},

$$\frac{1}{\alpha_b} = \frac{1}{3.95\times 10^{-3}}+(0-20) = \frac{1-(20\times 3.95\times 10^{-3})}{3.95\times 10^{-3}}.$$

$$\therefore \alpha_b = \frac{3.95\times 10^{-3}}{0.921} = 4.3\times 10^{-3}\ \text{deg C}^{-1}. \qquad \text{(b)}$$

(c) To obtain R_θ when $\theta = 160°$C, eqn. (5.10) may be used with the calculated value of α_{20}. Thus

$$\begin{aligned} R_{160} &= 76.0[1+3.95\times 10^{-3}(160-20)] \\ &= 76.0\times 1.55 = 118\ \Omega. \end{aligned} \qquad \text{(c)}$$

Alternatively, eqn. (5.9) and the values of R_0 and α_0 from part (b) may be used:

$$R_{160} = 70.0[1+(4.3\times10^{-3}\times160)]$$
$$= 70.0\times1.69 = 118\ \Omega. \quad \text{(c)}$$

5.24. The resistance of a resistor at a temperature of 25°C is 330 Ω. Calculate the resistance at (a) 75°C, and (b) 0°C. Assume a temperature coefficient of resistance of 0.0040 deg C^{-1} relative to 25°C. [(a) 400 Ω; (b) 300 Ω.]

5.25. A resistor has a resistance of 46.0 Ω at a temperature of 30.0°C and 47.3 Ω at 50.0°C. Calculate (a) the temperature coefficient of resistance at 30°C, and (b) the resistance at 100°C. [(a)1.4×10^{-3} deg C^{-1}; (b) 50.6 Ω.]

5.26. Given that the temperature coefficient of resistance of tungsten at a temperature of 20°C is 52×10^{-4} deg C^{-1}, calculate the coefficient at 0°C. [58×10^{-4} deg C^{-1}.]

5.27. A length of nickel wire has a resistance of 15 Ω at a temperature of 100°C. Calculate the resistance at 200°C, assuming the temperature coefficient of resistance to be 0.0054 deg C^{-1} relative to 0°C. Also calculate the error in taking the reference temperature of the coefficient to be 100°C. [20 Ω; +14%.]

Resistance Thermometry

5.28. Worked Example

A platinum resistance thermometer has a resistance of 23.0 Ω at a temperature of 0°C, 32.0 Ω at 100°C, and 61.0 Ω at 444.6°C on the ideal gas scale. Calculate the Celsius temperature (a) on the platinum resistance scale, and (b) on the ideal gas scale, when, in both cases, the resistance is 50.0 Ω.

Introduction

Resistance is not exactly a linear function of temperature on the ideal gas scale. Thus the equation

$$R_\theta = R_0(1+\alpha_0\theta)$$

is an over-simplification of the relationship between R_θ and θ, when θ is defined by the variation of pressure or volume of an ideal gas or by the variation of any property other than the resistance itself.

Allowance can be made for this by a power series:

$$R_\theta = R_0(1+a\theta+b\theta^2+c\theta^3 \ldots), \tag{5.14}$$

where a, b, and c are temperature coefficients; a is slightly different from α_0; b is less than a and may be negative; c is usually negligible; θ is the Celsius temperature on the ideal gas scale.

Alternatively, a different scale of temperature may be defined. Thus for platinum, which is the standard material for this purpose, a temperature scale is defined to have an exactly linear relationship with the platinum resistance, by the equation

$$R_\theta = R_0(1+\alpha\theta_p), \tag{5.15}$$

where θ_p is the temperature on this resistance scale. To agree with the ideal gas scale at the two fixed points of the Celsius system, i.e. 0°C and 100°C, α is defined by $R_{100} = R_0(1+100\alpha)$. Then eliminating α from eqn. (5.15) gives

$$\theta_p = 100(R_\theta - R_0)/(R_{100} - R_0). \tag{5.16}$$

Solution

(a) Using eqn. (5.16), $\theta_p = 100(50-23)/(32-23)$

$$= 300°\text{C}. \tag{a}$$

(b) The coefficients a and b are found by applying eqn. (5.14) to the data, neglecting c and coefficients of higher orders.

Thus at 100°C, $32.0 = 23.0(1+100a+10000b)$

and at 444.6°C, $61.0 = 23.0(1+444.6a+444.6^2\,b)$.

The solution of these simultaneous equations is found to be

$$a = 3.96\times10^{-3} \quad \text{and} \quad b = -5.7\times10^{-7}.$$

Then, at the unknown temperature,

$$50.0 = 23.0(1+3.96\times10^{-3}\theta-5.7\times10^{-7}\theta^2).$$

Now applying the standard formula for the solution of this quadratic gives two solutions:

$$\theta = 6640°\text{C} \quad \text{and} \quad \theta = 309°\text{C}.$$

The first of these temperatures is above the melting point of platinum (1773°C), but it appears as a solution because the quadratic form which has been assumed in eqn. (5.14) represents a parabola. Figure 5.1, which is not to scale, applies to the present case

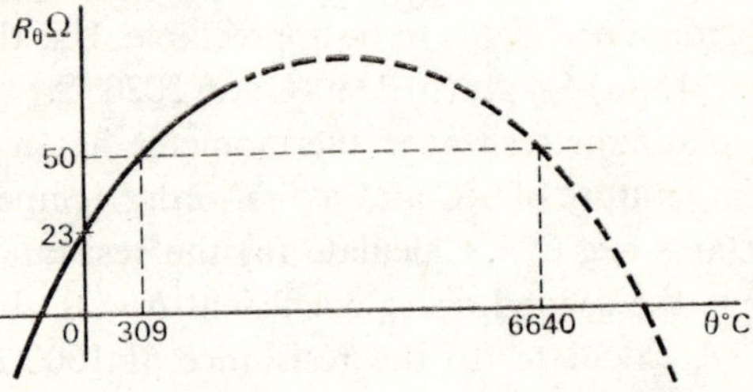

FIG. 5.1. Ex. 5.28.

where a is positive and b is negative. The equation and graph fail to take into account the change of state at the melting point, but are useful for temperatures below this point. Thus

$$\theta = 309°\text{C}. \tag{b}$$

5.29. The coil of a relay draws a steady current of 50 mA at a p.d. of 12 V when the temperature is 15°C. Calculate the temperature when the current drops to 45 mA at the same p.d. Assume the copper wire to have a linear variation of resistance with temperature. Ref.: copper. [42.5°C.]

5.30. A platinum resistance thermometer has a resistance of 10.0 Ω at a temperature of 0°C, and 13.9 Ω at 100°C. Calculate the Celsius temperature on the platinum resistance scale when the thermometer has a resistance of 21.0 Ω. [282°C.]

5.31. The tungsten filament of a lamp rated at 100 W, 200 V, has an operating temperature of 2200°C on the ideal gas scale. Calculate the initial current drawn by the filament at a temperature of 0°C when a p.d. of 200 V is applied. Assume the first- and second-order temperature coefficients of resistance for tungsten to be 5.2×10^{-3} deg C^{-1} and 7×10^{-7} deg C^{-2} respectively. [7.9 A.]

5.32. A metal filament lamp rated at 100 W, 240 V, has a resistance of 40.0 Ω at a temperature of 0°C. Calculate (a) the initial current when switched on at 20°C, (b) the operating temperature, assuming only the first-order temperature coefficient to be appreciable, and (c) the operating temperature, assuming both the first- and second-order coefficients to be appreciable. Use the coefficients given in ex. 5.31. [(a) 5.4 A; (b) 2580°C; (c) 2020°C.]

5.33. A platinum resistance thermometer has a resistance of 5.0 Ω at a temperature of 0°C and a first-order temperature coefficient $a = 0.0038$ deg C^{-1}. Calculate (a) the resistance at 1000°C, assuming that the second-order coefficient $b = 0$. If $b = -5.6 \times 10^{-7}$ deg C^{-2}, calculate (b) the resistance at 1000°C, and (c) the temperature for a resistance of 20 Ω. [(a) 24 Ω; (b) 21 Ω; (c) 910°C.]

5.34. A 240 V, 150 W lamp has a filament of resistance 30 Ω at 20°C. Calculate (a) the operating temperature, and (b) the change of temperature due to a drop of 10% in the supply p.d. Assume the resistance to vary linearly with the temperature at a rate of 0.50% per deg C and assume that the resistance at the operating temperature varies in proportion to the current required to maintain that temperature. [(a) 2600°C; (b) drop of 140°C.]

Thermistors

5.35. Worked Example

The resistance R of a thermistor is given by the expression $R = ae^{b/T}$, where a and b are constants, e is the base of natural logarithms, and T is the absolute temperature on the Kelvin (K) scale. If the resistance of a particular thermistor is 1.0 kΩ at a temperature of 20°C, and $b = 2600$ K, calculate the resistance at 100°C.

Solution

From the given expression, the resistances R_1 and R_2 at the temperatures T_1 and T_2 are given by

$$R_1 = ae^{b/T_1} \quad \text{and} \quad R_2 = ae^{b/T_2}.$$

$$\therefore R_1/R_2 = e^{b/T_1}/e^{b/T_2} = e^{b/T_1 - b/T_2}.$$

$$\therefore \log_e \frac{R_1}{R_2} = \frac{b}{T_1} - \frac{b}{T_2}.$$

Now if $T_1 = 20°\text{C} = 293$ K, $R_1 = 1000\ \Omega$;

$T_2 = 100°\text{C} = 373$ K, R_2 is required;

$b = 2600$.

Then $$\log_e \frac{1000}{R_2} = \frac{2600}{293} - \frac{2600}{373} = \frac{2600 \times 80}{293 \times 373} = 1.90.$$

From tables of Naperian logarithms it is seen that $1.90 = \log_e 6.7$.

$$\therefore \frac{1000}{R_2} = 6.7. \qquad \therefore R_2 = \frac{1000}{6.7} = 150\ \Omega.$$

5.36. A thermistor has a resistance of 500 Ω at a temperature of 0°C and 100 Ω at 30°C. Calculate (a) the resistance at 120°C, and (b) the temperature at which the resistance would be 10 Ω. [(a) 3.6 Ω; (b) 87°C.]

CHAPTER 6

ELECTROLYSIS

Electroplating

6.1. Worked Example

Calculate the current in a copper plating bath required to deposit a mass of 10 g of copper in a period of 1.0 h.

Introduction

When a metal, or a compound of a metal, dissolves to form a conducting solution, the solution contains ions (i.e. charged atoms or groups of atoms) including positively charged metal ions. If, then, the solution is made part of a circuit by immersing two electrodes, one connected to each terminal of a battery, the positive metal ions are attracted to the negative electrode (cathode). Here they become discharged by collecting electrons fed by the current in the external circuit, and may become deposited on the cathode as a surface layer of metal plate.

If each metal ion has the same mass and charge, the total mass of the metal deposited is proportional to the total charge supplied by the current. This is a particular example of Faraday's first law of electrolysis, which extends the same relationship to other substances in addition to metals:

The mass M of a substance liberated in electrolysis is proportional to the charge Q passed by the current.

Then
$$M = zQ, \tag{6.1}$$

where z = the constant of proportionality, called the electrochemical equivalent (e.c.e.) of the particular substance.

Solution

From the reference data it is seen that the e.c.e. of copper $z = 3.3 \times 10^{-7}$ kg C^{-1}. But the mass M deposited = 10 g = 10^{-2} kg.

$$\therefore\ Q = \frac{M}{z} = \frac{10^{-2}}{3.3 \times 10^{-7}} = 3.03 \times 10^{4}\ \text{C}.$$

Now the current I must pass this charge in the time t = 1 h = 3600 s. So

$$I = \frac{Q}{t} = \frac{3.03 \times 10^{4}}{3600} = 8.4\ \text{A}.$$

6.2. A mass of 0.791 g of copper is deposited electrolytically by a current of 2.00 A passing for a period of 20.0 min. Calculate the electrochemical equivalent of copper. [3.3×10^{-7} kg C^{-1}.]

6.3. It is required to silver plate an area of 100 cm^2 to a depth of 0.10 mm in a time of 1.0 h. Calculate the current required. Ref.: e.c.e. and density of silver. [2.6 A.]

6.4. A silver voltameter and a copper voltameter are connected in series with a d.c. supply and an ammeter indicating a current of 3.0 A. After a period of 30.0 min the silver cathode is found to have increased in mass by 5.81 g and the copper by 1.71 g. Assuming the e.c.e. of silver given in the reference data, calculate (a) the e.c.e. of copper, and (b) the error of the ammeter. [(a) 3.3×10^{-7} kg C^{-1}; (b) +3.9%.]

Voltaic Cells

6.5. Worked Example

Calculate the mass of lead sulphate formed per ampere-hour in a single cell of a lead acid accumulator on discharge. The chemical formula for lead sulphate is $PbSO_4$; the atomic weights of Pb, S, and O are 207, 32, and 16 respectively; the e.c.e. of lead is 1.07×10^{-6} kg C^{-1}.

Solution

It is assumed that lead sulphate is formed at both negative and positive plates in equal amounts when the accumulator provides a current. Also, although the lead is not liberated from the solution, the quantity of lead which combines with the liberated sulphate ions is given by the e.c.e. of lead.

Now 1 ampere-hour (Ah) is the charge passed by a current of 1 A flowing for a time of 1 h or 3600 s; i.e. a charge of 3600 C.

Then, using eqn. (6.1), the mass of lead reacting at each plate = $M = zQ = 1.07 \times 10^{-6} \times 3600$ kg.

So for both positive and negative plates,

$$2M = 2 \times 1.07 \times 10^{-6} \times 3600 \text{ kg} = 7.7 \text{ g}.$$

Now the molecular weight of $PbSO_4$ is the sum of the weights of the constituent atoms = $207 + 32 + (4 \times 16) = 303$ units. Thus 303 units of $PbSO_4$ are formed for each 207 units of Pb in the reactions.

Therefore in a discharge of 1 Ah the mass of $PbSO_4$ formed from 7.7 g of Pb is

$$\frac{303}{207} \times 7.7 \text{ g} = 11 \text{ g}.$$

6.6. A dry Leclanché battery has three cells in series each having an e.m.f. of 1.5 V and negligible internal resistance. Calculate the mass of zinc chloride ($ZnCl_2$) formed in the battery when it is connected to a circuit of resistance 30 Ω for 30 min. The e.c.e. of zinc is 3.39×10^{-7} kg C^{-1}, and the atomic weights of zinc and chlorine are 65.4 and 35.5 respectively. [570 mg.]

6.7. A 12 V accumulator has a capacity of 40 Ah. Calculate the time for which it can deliver (a) a current of 8.0 A, and (b) a power of 8.0 W. Also calculate (c) the total charge which it can circulate, and (d) the total energy which it can deliver. [(a) 5 h; (b) 60 h; (c) 1.4×10^5 C; (d) 1.7×10^6 J.]

6.8. A nickel–iron battery gives a discharge of 100 Ah at an average p.d. of 1.2 V. It is fully recharged in a period of 9.0 h at an average p.d. of 1.5 V and current of 15 A. Calculate the efficiency as a comparison of input and output of (a) charge, and (b) energy. [(a) 74%; (b) 59%.]

Faraday's Second Law

6.9. Worked Example

From the following data, calculate (a) the e.c.e. of oxygen, and (b) Faraday's constant. For oxygen, atomic weight = 16 and valency = 2. For hydrogen, atomic weight = 1.008 and e.c.e. $= 1.045 \times 10^{-8}$ kg C^{-1}.

Introduction and Solution

An ion is an atom or group of atoms which has an unbalanced electric charge due to the loss or gain of one electron or more. The number of electrons gained or lost equals the valency of the ion. Thus an ion of oxygen of valency 2 carries an unbalanced negative charge q_O of two electrons, whereas an ion of hydrogen has an unbalanced positive charge due to the loss of a negative charge q_H of one electron.

$$\therefore\ q_O = 2q_H.$$

The mass m_O of an oxygen atom and the mass m_H of a hydrogen atom are in proportion to their atomic weights (relative atomic masses). So $m_O = (16/1.008)m_H$.

Now if N atoms of one kind are liberated by the neutralization of the charges of N ions, then the total mass $M = Nm$, and the total charge $Q = Nq$. But M/Q is the electrochemical equivalent z.

$$\therefore\ z = M/Q = (NM)/(Nq) = m/q. \tag{6.2}$$

Thus for oxygen, $z_O = m_O/q_O = (16/1.008)m_H/2q_H$.

But $m_H/q_H = z_H$ – c.c.c. of hydrogen.

$$z_O = (8/1.008)z_H \tag{6.3}$$

$$= (8/1.008)\times 1.045\times 10^{-8} = 8.3\times 10^{-8}\,\text{kg C}^{-1}. \qquad \text{(a)}$$

Alternatively, eqn. (6.3) may be stated as an instance of Faraday's second law of electrolysis:

The e.c.e. of a substance is proportional to its chemical equivalent.

The chemical equivalent E of an element is the ratio of the atomic weight to the valency. Thus $E = 16/2$ for oxygen and $1.008/1$ for hydrogen.

(b) From Faraday's second law just referred to it follows that $E = Fz$, where F may be called Faraday's constant, and is the same for all substances. Using the figures for hydrogen given in the question, $F = E_H/z_H = 1.008/1.045\times 10^{-8}$

$$= 9.65\times 10^{7}\,\text{C kg}^{-1}\ \text{equivalent}. \qquad \text{(b)}$$

Similarly for oxygen, $E_O/z_O = 8/8.3\times 10^{-8} = 9.65\times 10^{7}$ (again).

The units stated for F are derived from the expression obtained by substituting M/Q for z. Thus $F = E/z = Q/(M/E)$, where, in M.K.S. units, Q is in coulombs, and M/E is the mass expressed as a multiple of the equivalent weight in kilograms, i.e. the number of kilogram equivalents.

6.10. A mass of 2.09 g of a certain divalent metal is deposited in electrolysis by a current of 3.00 A for 20.0 min. Calculate the atomic weight of the metal assuming Faraday's constant of electrolysis. [112.]

6.11. When a current of 2.5 A is passed through a solution of a lead salt for a time of 20 min, a mass of 3.21 g of the metal is deposited. Assuming the atomic weight of lead to be 207, and assuming the value of Faraday's constant of electrolysis, calculate the valency of lead. [2.]

6.12. A mass of 20 mg of hydrogen is liberated in the electrolysis of acidulated water, using a current of 5.0 A. Calculate (a) the time taken, and (b) the volume of water decomposed. The atomic weights of hydrogen and oxygen are 1.008 and 16 respectively. Ref.: Faraday's constant of electrolysis and density of water. [(a) 6.4 min; (b) 0.18 cm^3.]

Avogadro's Number

6.13. Worked Example

During the electrolysis of dilute acid, a mass of 47.0 mg of hydrogen is liberated by a current of 2.50 A passing for a time of 30.0 min. Assuming the charge of each hydrogen ion to be due to the loss of one electron of charge 1.60×10^{-19} C, calculate (a) the mass of one atom of hydrogen, (b) the total charge carried by 1 kg atom of hydrogen, and (c) the number of atoms in 1 kg atom of any element.

Solution

(a) Using eqn. (6.2), the mass of a hydrogen atom $m_H = q_H z_H$, where q_H is the given charge of one electron, and z_H is given by M/Q,

Now $M = 47\times10^{-6}$ kg and $Q = It$, where I is the current of 2.5 A and t is the time of 1800 s.

$$\therefore\ z_H = M/Q = 47\times10^{-6}/(2.5\times1800) = 1.045\times10^{-8}\ \text{kg C}^{-1}, \tag{6.4}$$

and the mass of one atom of hydrogen $m_H = q_H z_H$

$$= 1.6\times10^{-19}\times1.045\times10^{-8} = 1.67\times10^{-27}\text{ kg.} \qquad \text{(a)}$$

(b) One kilogram atom of an element is the mass A_r in kilograms, where A_r equals the relative atomic mass or "atomic weight". Therefore one kilogram atom of hydrogen has a mass of 1.008 kg, and, from eqn. (6.4), the charge Q carried by this mass

$$= 1.008/(1.045\times10^{-8}) = 9.65\times10^{7}\text{ C.} \qquad \text{(b)}$$

This is numerically equal to Faraday's constant, as in ex. 6.9, since the chemical equivalent and the atomic weight are equal for hydrogen.

(c) In terms of the symbols already defined, the mass of a single atom of any element is given by $A_r m_H$, and therefore the number of atoms in a kilogram atom of the element equals

$$A_r/(A_r m_H) = 1/(1.67\times10^{-27}) = 6.0\times10^{26}\text{ atoms/kg atom.} \qquad \text{(c)}$$

This is the same for all elements whatever A_r may be, and also equals the number of molecules per kilomole (kilogram molecule) for which the relative molecular mass M_r would replace A_r in the above equation. This is Avogadro's number in M.K.S. units.

6.14. A volume of 10 cm³ of hydrogen at a temperature of 20°C and pressure of 780 mmHg is produced in electrolysis by a current of 0.50 A for a period of $2\frac{3}{4}$ min. Calculate the charge of a hydrogen ion. Assume the value of Avogadro's number and assume that each ion is a single charged atom of atomic weight 1.008, and that the density of hydrogen is 0.090 kg m^{-3} at s.t.p. [1.6×10^{-19} C.]

6.15. A current of 2.0 A is used in the electrolysis of acidulated water. Calculate the time taken to liberate (a) 1.0 l of oxygen at s.t.p., and (b) 10^{12} molecules of oxygen. Assume Avogadro's number and the charge of the electron as given in the reference data. Also assume oxygen to have diatomic molecules, atomic weight of 16, density of 1.43 kg m^{-3} at s.t.p., and valency of 2. [(a) 2.4 h; (b) 0.32 μs.]

CHAPTER 7

THERMOELECTRICITY

Thermocouple E.M.F.

7.1. Worked Example

A thermocouple with one junction at a temperature of 0°C gives an e.m.f. of 0.90 mV when the other junction is at a temperature of 100°C, and 0.50 mV at 50°C. Calculate (a) the two possible temperatures of the hot junction for which the e.m.f. is 1.00 mV when the cold junction is at 0°C, (b) the neutral temperature, and (c) the e.m.f. when the hot junction is at 100°C and the cold junction at 20°C.

Introduction

The e.m.f. of a thermocouple varies with the temperature of one junction when the other is at constant temperature. If temperatures are measured on the ideal gas scale, or more simply on a mercury-in-glass thermometer calibrated in the ordinary way, the e.m.f. is not linearly related to the temperature. The e.m.f. is given by a power series

$$E = (A\theta + B\theta^2 + C\theta^3 + \ldots) - (A\theta_0 + B\theta_0^2 + C\theta_0^3 + \ldots), \quad (7.1)$$

where θ is the temperature of the "hot" junction, θ_0 is that of the "cold" junction, and A, B, C, etc., are characteristic constants depending on the particular metals of the thermocouple and the temperature scale. Unless otherwise stated it is assumed that the coefficients of the third and higher powers of θ and θ_0 are negligible. Then

$$E = A(\theta - \theta_0) + B(\theta^2 - \theta_0^2). \tag{7.2}$$

Also, if the Celsius scale is used and the cold junction is held at the ice point, $\theta_0 = 0°C$, and then

$$E = A\theta + B\theta^2. \tag{7.3}$$

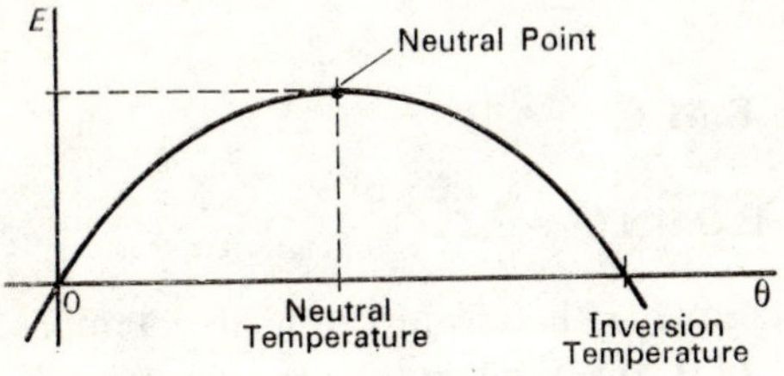

FIG. 7.1. Temperature–e.m.f. curve for a thermocouple (ex. 7.1).

The graph of eqn. (7.3) is a parabola, and is shown in Fig. 7.1 for the case where A is positive and B is negative. For a perfect parabola, the neutral temperature, at which the e.m.f. reaches its maximum, is half the temperature of the inversion point, at which the e.m.f. reverses its direction.

The signs of A and B depend upon a sign convention used in comparing the e.m.f.s obtained by coupling various metals in turn with a common metal, usually lead or platinum. Thus, when metal X is coupled with lead, E and A are positive if the e.m.f. acts across the hot junction from lead to X when $\theta_0 = 0°C$ and θ is a small positive temperature. B is positive if this e.m.f. rises more rapidly than in direct proportion to θ.

Solution

(a) Now substituting the data in eqn. (7.3) gives two equations to be solved simultaneously:

$$0.90 = 100A + 10000B, \tag{7.4}$$

$$0.50 = 50A + 2500B. \tag{7.5}$$

Then, (7.5)×2:

$$1.00 = 100A + 5000B.$$

Subtracting eqn. (7.4):

$$0.10 = 0 - 5000B.$$

$$\therefore\ B = \frac{-0.10}{5000} = -2 \times 10^{-5}\ \text{mV deg C}^{-2}.$$

Then, from eqn. (7.4) $\quad 100A = 0.90 - 10000B.$

$$\therefore\ A = 0.0090 - 100(-2 \times 10^{-5}).$$

$$= 11 \times 10^{-3}\ \text{mV deg C}^{-1}.$$

Now if θ_1 is the temperature at which $E = 1.00$ mV, eqn. (7.3) with the numerical values of A and B gives

$$1.0 = (11 \times 10^{-3}\theta_1) - (2 \times 10^{-5}\theta_1^2) \quad \text{or} \quad 2\theta_1^2 - 1100\theta_1 + 10^5 = 0.$$

This equation is solved by the usual formula:

$$\frac{1100 \pm \sqrt{[1100^2 - (4 \times 2 \times 10^5)]}}{4} = 275 \pm \frac{\sqrt{(41 \times 10^4)}}{4}$$

$$= 275 \pm \frac{640}{4} = 440°\text{C} \quad \text{or} \quad 120°\text{C} \quad \text{(2 sig. fig.)}. \qquad \text{(a)}$$

(b) From the symmetry of the parabola, the temperature of the neutral point is seen to be the arithmetic mean of the values for (a).

$$\therefore\ \text{the neutral temperature} = \frac{440 + 120}{2} = 280°\text{C}. \qquad \text{(b)}$$

(c) When $\theta_0 = 20°$C, eqn. (7.2) applies: i.e. the cold junction correction $(A\theta_0 + B\theta_0^2)$ is to be subtracted from $(A\theta + B\theta^2)$, which is known to be 0.90 mV for $\theta = 100°$C. Now

$$A\theta_0 + B\theta_0^2 = (11 \times 10^{-3} \times 20) - (2 \times 10^{-5} \times 400) = 21.2 \times 10^{-2}\ \text{mV}.$$

Therefore the e.m.f. with cold junction correction is

$$0.90 - 0.212 = 0.69\ \text{mV}. \qquad \text{(c)}$$

7.2. A copper–iron thermocouple has characteristic constants $A = 14.2\ \mu V\ \text{deg}\ C^{-1}$ and $B = -0.026\ \mu V\ \text{deg}\ C^{-2}$. Calculate the e.m.f. when the cold junction is at 50°C and the hot junction at 275°C. [1.3 mV.]

7.3. The two metals of a thermocouple are pure platinum and an alloy of 10% rhodium 90% platinum. When the cold junction is at 0°C, the e.m.f. is 0.642 mV for a hot junction temperature of 100°C and 0.430 mV for an unknown temperature. A platinum resistance thermometer at the same temperatures has resistance of 25.0 Ω at 0°C, 34.8 Ω at 100°C, and 31.9 Ω at the unknown temperature. Calculate the unknown temperature (a) on the resistance scale of temperature, and (b) on the thermocouple scale of temperature, assuming a linear scale in each case. [(a) 70.4°C; (b) 67.0°C.]

7.4. A copper–iron thermocouple with one junction at 0°C has a neutral point at 275°C, where the e.m.f. is 1.80 mV. Calculate (a) the inversion temperature, (b) the e.m.f. at 150°C, and (c) the other temperature at which the e.m.f. is equal to that at 150°C. [(a) 550°C; (b) 1.43 mV; (c) 400°C.]

7.5. The e.m.f. of a thermocouple with its cold junction at 0°C is 2.39 mV when the hot junction is at 300°C and 5.57 mV at 600°C. Calculate the e.m.f. at 400°C, assuming that the variation of e.m.f. with temperature is (a) linear between 300°C and 600°C, and (b) quadratic between 0°C and 600°C. [(a) 3.45 mV; (b) 3.36 mV.]

7.6. A chromel–alumel thermocouple, with one junction at 0°C, gives an e.m.f. of 16.4 mV when the other junction is at 400°C, and 33.3 mV at 800°C. Calculate (a) the e.m.f. at 1200°C, and (b) the temperature at which the e.m.f. is 40 mV. [(a) 50.7 mV; (b) approximately 950°C.]

7.7. A thermocouple with cold junction at 0°C has an e.m.f. of 0.96 mV when the hot junction is at 100°C, and 1.32 mV at the neutral point which is above 100°C. Calculate the temperature of the neutral point. [209°C.]

Thermoelectric Power

7.8. Worked Example

The characteristic constants for two thermocouples are as follows: for iron–lead, $A = 17\ \mu V\ \text{deg}\ C^{-1}$ and $B = -0.020\ \mu V\ \text{deg}\ C^{-2}$; and for copper–lead, $A = 2.8\ \mu V\ \text{deg}\ C^{-1}$ and $B = 0.006\ \mu V\ \text{deg}\ C^{-2}$. Draw the thermoelectric diagram for these couples and hence obtain the e.m.f. of a copper–iron thermocouple when the two junction temperatures are (a) 100°C and 200°C, and (b) 200°C and 300°C.

Introduction

The thermoelectric power of a thermocouple is a measure of its sensitivity to temperature change, and is defined as the rate of change of e.m.f. with temperature, usually in $\mu V\ \text{deg}\ C^{-1}$.

Now eqn. (7.3) gives $E = A\theta + B\theta^2$.

Then, by differentiation, the thermoelectric power $dE/d\theta = A + 2B\theta$.

The thermoelectric diagram is a graph of $dE/d\theta$ against θ (Fig. 7.2).

Now

$$\int_0^{\theta_1} \frac{dE}{d\theta}\, d\theta = E_1.$$

Therefore the area under the graph represents the e.m.f. for the temperatures shown by the base.

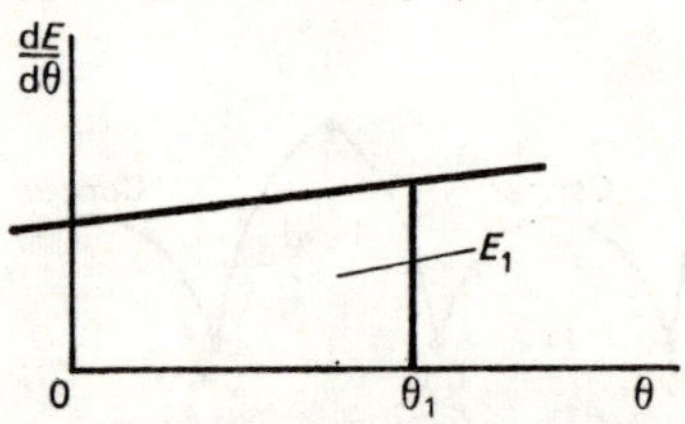

FIG. 7.2. Thermoelectric diagram (ex. 7.8).

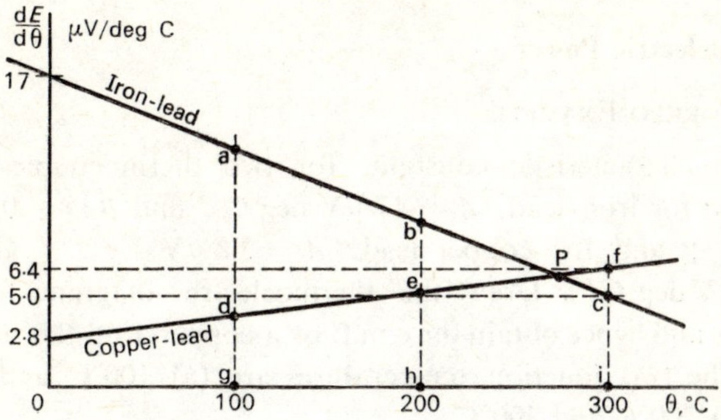

FIG. 7.3. Graphical solution to ex. 7.8.

For the iron–lead thermocouple the power

$$dE/d\theta = A+2B\theta = 17-0.040\theta\ \mu\text{V deg C}^{-1}.$$

This line runs through the points ($\theta = 0$, $dE/d\theta = 17$) and ($\theta = 300$, $dE/d\theta = 5$), as shown in Fig. 7.3. Similarly, for the copper–lead thermocouple the line $dE/d\theta = 2.8+0.012\theta$ runs through the points ($\theta = 0$, $dE/d\theta = 2.8$) and ($\theta = 300$, $dE/d\theta = 6.4$).

Further Introduction

If two copper–lead thermocouples are connected in series opposition as in Fig. 7.4, the two junctions at 0°C give equal and opposite e.m.f.s, and the resultant e.m.f. is due only to the other junctions. Hence the two couples in series opposition are equivalent to one couple with junctions at 100°C and 200°C.

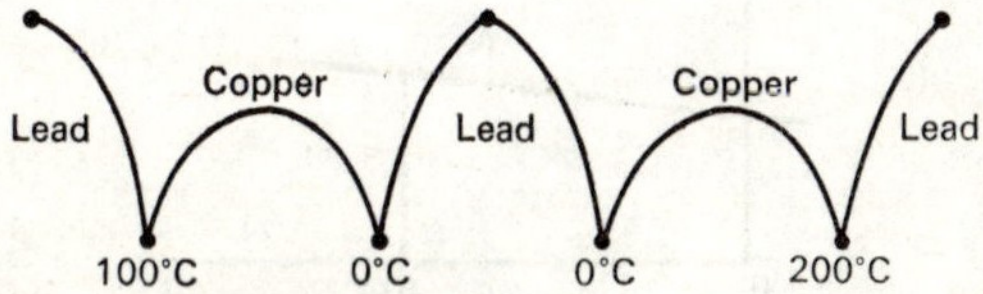

FIG. 7.4. Thermocouples in series opposition.

The resultant e.m.f. is represented in the thermoelectric diagram (Fig. 7.3) by the area *dehg* which is the difference of the areas representing the e.m.f.s of the separate couples.

Similarly, a copper–lead couple and an iron–lead couple in series opposition may be shown to be equivalent to a copper–iron couple.

Solution

(a) For junction temperatures of 100°C and 200°C, the e.m.f. of the copper–iron couple is represented by the difference of the areas *abhg* and *dehg*, i.e. *abed*. Now the points *a*, *b*, *e*, *d* have ordinates 13, 9, 5.2 and 4 respectively.

$$\text{So the area } abed = 100\left(\frac{13+9}{2}\right)-100\left(\frac{4+5.2}{2}\right)$$

$$= 1100-460. \qquad \therefore \text{ e.m.f.} = 640\ \mu\text{V}. \qquad \text{(a)}$$

(b) For temperatures of 200°C and 300°C, the e.m.f. is represented by the difference of areas *bcih* and *efih*, extending both sides of the neutral point *P*. Area *bPe* represents positive e.m.f. and area *Pcf* represents negative e.m.f.

$$\text{Now } bcih-efih = 100\left(\frac{9+5}{2}\right)-100\left(\frac{5.2+6.4}{2}\right)$$

$$= 700-580. \qquad \therefore \text{ e.m.f.} = 120\ \mu\text{V}. \qquad \text{(b)}$$

Note that the e.m.f.s could be obtained from eqn. (7.2) of ex. 7.1 in which the constants A and B are the differences of those of iron–lead and copper–lead.

Thus, for copper–iron,

$$E = (17-2.8)(\theta-\theta_0)+(-0.02-0.006)(\theta^2-\theta_0^2).$$

7.9. Given the characteristic constants of ex. 7.8, calculate (a) the e.m.f. and temperature at the neutral point, and (b) the inversion temperature, both for a thermocouple of iron and copper with the cold junction at 0°C. [(a) 1.9 mV, 270°C; (b) 540°C.]

7.10. A certain thermocouple has a thermoelectric power of 16 μV deg C^{-1} at a temperature of 0°C and 30 μV deg C^{-1} at 250°C. Calculate the e.m.f. at 250°C (a) from the thermoelectric diagram, and (b) from the e.m.f. equation. Assume a cold junction temperature of 0°C. [5.8 mV.]

7.11. The thermoelectric power of a copper–constantan thermocouple is given as $41+0.10\theta$ μV deg C^{-1}, and that of a copper–nickel thermocouple as $19+0.070\theta$ μV deg C^{-1}. Calculate the e.m.f. and thermoelectric power of a nickel–constantan thermocouple with junction temperatures of 0°C and 100°C. [2.4 mV; 25 μV deg C^{-1}.]

7.12. A thermocouple consists of three wires of total resistance 1.5 Ω connected in series across the terminals of a millivoltmeter of resistance 12 Ω. Two wires are of the same material, while the third wire is of different material and is connected between the others with junction temperatures of 0°C and 200°C. The characteristic constants of the thermocouple are $A = +16$ μV deg C^{-1} and $B = +0.005$ μV deg C^{-2}. Calculate (a) the e.m.f. and thermoelectric power, and (b) the p.d. and electrical power supplied to the millivoltmeter. [(a) 3.4 mV, 18 μV deg C^{-1}; (b) 3.0 mV, 0.76 μW.]

CHAPTER 8

D.C. GENERATORS

Flux Cutting

8.1. Worked Example

A conductor of length 8.0 cm lies perpendicular to a magnetic field of uniform flux density 0.40 T. Calculate the e.m.f. induced in the conductor (a) when it moves at a rate of 20 m s^{-1} in a direction perpendicular to its length and to the flux, (b) when it moves at a rate of 20 m s^{-1} perpendicular to its length and at an angle of 60° to the flux, and (c) when it rotates about one end, like a spoke of a wheel, at a rate of 5000 rev min^{-1} in a plane perpendicular to the flux.

Introduction

Figure 8.1 shows a conductor of length l moving with velocity v perpendicular to itself and to the flux between magnet poles N and S. As the conductor cuts across the flux the deflection of the galvanometer G would indicate a current I produced by an e.m.f. E induced in the circuit. The electrical power is produced by conversion from the mechanical power required to drive the conductor by an external force F. Therefore if no other energy is concerned, the mechanical input power equals the electrical output power.

Thus

$$Fv = EI, \tag{8.1}$$

where F is in newtons, v in metres per second, E in volts, I in amps, making the power in watts.

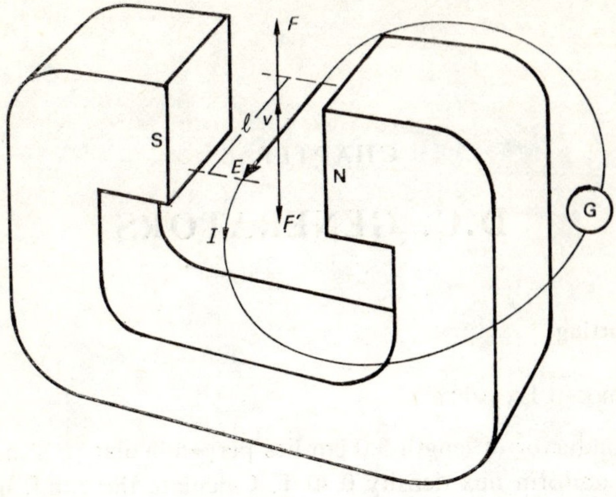

FIG. 8.1. EX. 8.1.

The applied force F is required to balance the reaction force F', given by eqn. (1.6). Thus $F = F' = BIl$, where B is assumed to be uniform across the length l and negligible elsewhere.

$$\therefore\ (BIl)v = EI. \qquad \therefore\ E = Bvl, \tag{8.2}$$

where B, v, and l are perpendicular to each other.

Now the flux density B is the flux per unit area and the area swept by the moving conductor per second is vl. Bvl is the rate at which flux is cut in webers per second.

Thus the e.m.f. equals the rate at which flux is cut by the conductor. (8.3)

This last statement applies in general, even if B, v, and l are not perpendicular to each other.

Solution

(a) In the particular question, eqn. (8.2) gives

$$E = Bvl = 0.4 \times 20 \times (8 \times 10^{-2}) = 0.64\ \text{V}. \tag{a}$$

(b) When the velocity is 20 m s^{-1} at an angle of 60° to the direction of the flux, the effective velocity at which the conductor cuts across the flux is 20 sin 60° (Fig. 8.2). Then

$$E = 0.64 \times \sin 50° = 0.64 \times 0.866 = 0.55 \text{ V}. \qquad \text{(b)}$$

Alternatively, it may be noticed that whereas the area swept in unit time is vl, the flux which is cut in this case (b) depends on the projection of this area perpendicular to the flux, i.e. $vl \sin 60°$.

(c) In one revolution the 8 cm length sweeps a circular area $A = \pi \times (8 \times 10^{-2})^2 = 0.0201 \text{ m}^2$. The flux through this area is BA, and this is cut n times per second, where n is the frequency of rotation in rev s^{-1}. Then from statement (8.3),

$$E = BAn = 0.4 \times 0.0201 \times \frac{5000}{60} = 0.67 \text{ V}. \qquad \text{(c)}$$

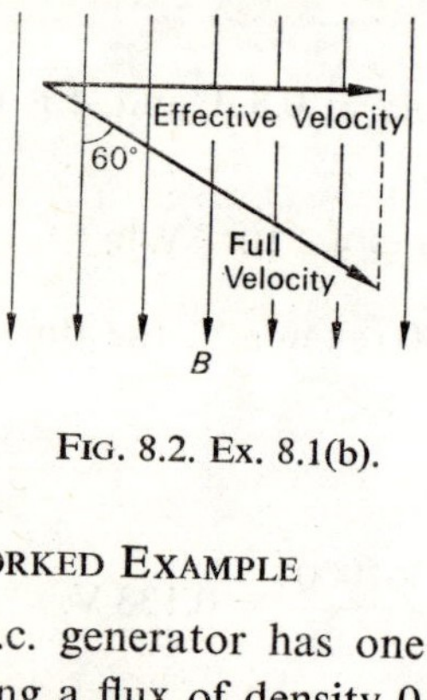

FIG. 8.2. Ex. 8.1(b).

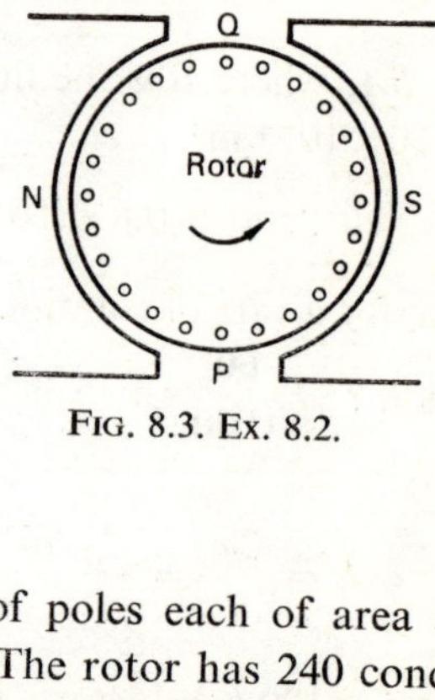

FIG. 8.3. Ex. 8.2.

8.2. WORKED EXAMPLE

A d.c. generator has one pair of poles each of area 50 cm^2 producing a flux of density 0.80 T. The rotor has 240 conductors connected in two parallel paths, each having 120 conductors in series. Calculate the output e.m.f. when the rotor turns at a rate of 1000 rev min^{-1}.

Introduction

Consider a conductor, shown by its small circular cross-section at P in Fig. 8.3, passing between one pole face and the other as the rotor turns. In the next complete rotation, the e.m.f. generated in this conductor is first in one direction along its length, as it passes across the face of pole S, and then the opposite across N. The operation of the commutator (not shown) ensures that this e.m.f., and those of all the other conductors, are led to the output terminals, always in the same direction, with half of them in series in each of two parallel paths.

Now, as stated in ex. 8.1, the e.m.f. in a conductor equals the rate at which the flux is cut by it. This rate is constant as the conductor passes across each pole face, but drops to zero in passing from one pole to the other.

The mean e.m.f. E in each conductor is then $2\varphi/t$, where φ is the flux in webers passing out of the N face and into the S, and t is the time in seconds for one complete rotation past both poles.

Solution

$\varphi = BA$, where B is the flux density $= 0.8$ T and A is the face area $= 50\times 10^{-4}$ m^2.

$$\therefore \varphi = 0.8\times(50\times 10^{-4}) = 4\times 10^{-3}\text{ Wb}.$$

Also, for a rate of rotation of 1000 rev min^{-1}, the time for one revolution $t = \dfrac{60}{1000}$ s.

Then

$$E = \frac{2\varphi}{t} = \frac{2\times(4\times 10^{-3})\times 1000}{60} = 0.133\text{ V}.$$

Now for 120 conductors, each with the same e.m.f., in series in each parallel path, the resultant e.m.f. is

$$120E = 120\times 0.133 = 16\text{ V}.$$

8.3. The rotor conductors of a d.c. generator are 15 cm in length, parallel to the shaft, and move at a speed of 3000 cm s^{-1}. Each stator pole has a face area of 10×15 cm^2, and a magnetic flux of 12 mWb, assumed to be of constant density and to be completely cut by the rotor conductors. Calculate the maximum e.m.f. generated in each conductor. [3.6 V.]

8.4. A flat coil of 40 turns, 2.0 cm in diameter, lies with its plane perpendicular to a magnetic field. Its plane is turned through 180° in a time of 10 ms. Calculate the flux density of the field if the induced e.m.f. has an average value of 0.15 V. [0.060 T.]

8.5. Calculate the speed in kilometres per hour with which a conductor of length 30 cm moves in a d.c. generator when the e.m.f. induced in it is 5.0 V as it passes a pole giving a flux density of 0.50 T across the full length of the conductor. [120 km h^{-1}.]

8.6. Calculate the e.m.f. induced in an axle of length 2.0 m as the vehicle to which it is fitted travels at a speed of 80 km h^{-1}. The flux density of the earth's magnetic field is to be taken as 50 μT with an angle of dip of 70°. [2.1 mV.]

8.7. The flux per pole of a two-pole generator is 10^{-2} Wb and the rotor has 150 conductors, half of them in series in each of two parallel paths. Calculate the rate of rotation at which the generated e.m.f. would be 50 V. [2000 rev min^{-2}.]

8.8. A four-pole d.c. generator has a rotor of 120 conductors, half of them in series in each of two parallel paths. The magnetic flux per pole is 20 mWb. Calculate the e.m.f. generated at a speed of 1500 rev min^{-1}. [120 V.]

Torque and Power

8.9. WORKED EXAMPLE

A simple two-pole generator has magnetic flux of mean density 0.60 T through the cylindrical surface of the rotor. The rotor carries a single square coil of 25 turns, 10×10 cm. Calculate the armature torque and power required to maintain a speed of 2000 rev min^{-1} with an efficiency of 88% when the current in the coil is 5.0 A.

Introduction and Solution

The force F on each side of length l of each turn of the coil when current I is flowing, is given by $F = BIl$ [see eqn. (1.6)]. Since the flux density B is less at the edge of a pole face than at the centre, the mean value of B is used to give the mean value of the force F. Thus

$$F = 0.6 \times 5 \times 0.1 = 0.3 \text{ N.}$$

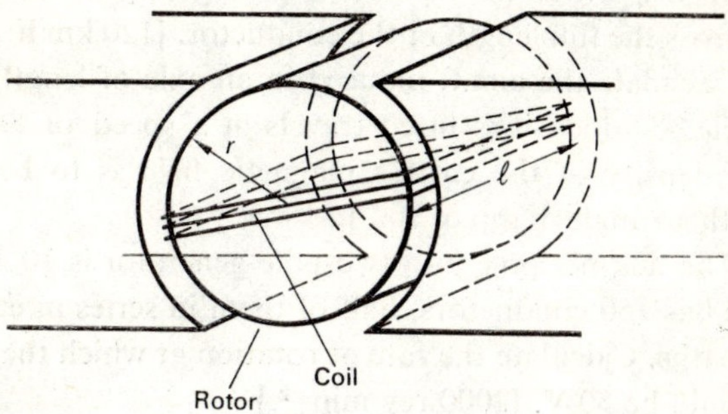

FIG. 8.4. Ex. 8.9.

The turning moment of this force about the rotor shaft is Fr, where r is the radius of the circle in which the coil rotates, = 0.05 m (Fig. 8.4). Thus for two sides of each of the 25 turns of the coil, the resultant mean torque on the rotor or armature

$$50 \times 0.3 \times 0.05 = 0.75 \text{ newton metres (Nm).}$$

This torque could be calculated from eqn. (1.7) of ex. 1.7, which applies to the somewhat similar arrangement of the pivoted coil in a moving-coil instrument.

In the generator, this torque is one of reaction against the forward rotation of the rotor and must be overcome by the torque applied externally. Also, due to friction, air resistance, and other losses, giving an efficiency of 88% in this case, the required torque is greater than that above, in the ratio 100 : 88.

Then the required torque

$$T = 0.75 \times \frac{100}{88} = 0.85 \text{ Nm}.$$

Now power $P = T\omega$, where ω = angular velocity in rad s^{-1}. Then the required power $= 0.85 \times (2000 \times 2\pi/60) = 180$ W.

Alternatively, the required power may be calculated from the output power $P = EI$, where I is the output current of 5 A and E is the output e.m.f. for both sides of the 25 turns, i.e. 50 Bvl. Here the velocity for 2000 rev min^{-1} is $v = 2000/60 \times \pi \times 0.1 = 10.5$ m s^{-1}.

Then the output power

$$P = 50\ BvlI = 50 \times 0.6 \times 10.5 \times 0.1 \times 5 = 157 \text{ W},$$

and the input power $= 157 \times 100/88 = 180$ W.

8.10. A conductor of length 20 cm lies perpendicularly across a magnetic field of flux density 0.30 T and forms part of a circuit of resistance 3.0 Ω. Calculate the force and power required to move the conductor at a speed of 700 cm s^{-1} perpendicular to the flux and to the length of the conductor. [0.0084 N; 60 mW.]

8.11. A conductor, 30 cm long, forming part of a closed circuit of 50 Ω resistance, lies in a direction perpendicular to a field of flux density 0.80 T. Calculate the force and speed with which the conductor must be moved in a direction perpendicular to both its length and the flux to generate a power of 0.50 W. [0.024 N; 21 m s^{-1}.]

8.12. A two-pole generator has a rotor of diameter 10 cm and length 15 cm, carrying 80 conductors, half of them in each of two parallel paths. The flux per pole is 12 mWb. Calculate (a) the mean density of flux over the cylindrical surface of the rotor, and (b) the torque required to generate an output power of 10 W in a circuit of resistance 10 Ω, assuming an efficiency of 80%. [(a) 0.51 T; (b) 0.19 Nm.]

Shunt Generators

8.13. Worked Example

A shunt generator has an armature resistance of 0.05 Ω and a field resistance of 50 Ω. It provides an output current of 20 A at a p.d. of 110 V with an efficiency of 81%. Calculate (a) the field losses, (b) the armature losses, (c) the mechanical input power, and (d) all losses other than those of (a) and (b).

Introduction

In a shunt generator the stator poles are magnetized by field windings which are supplied with current driven by the e.m.f. generated in the armature. Thus in Fig. 8.5 the field windings, of resistance R_f, are connected in parallel with the armature of resistance R_a. Then the e.m.f. E, generated in the armature, drives through the armature a total current I_a which divides into I_f, through the field windings, and I_l through the load R_l connected to the output terminals.

Then

$$I_a = I_f + I_l. \tag{8.4}$$

The p.d. V between these terminals is less than the e.m.f. E, due to the internal resistance of the armature R_a. Thus for the

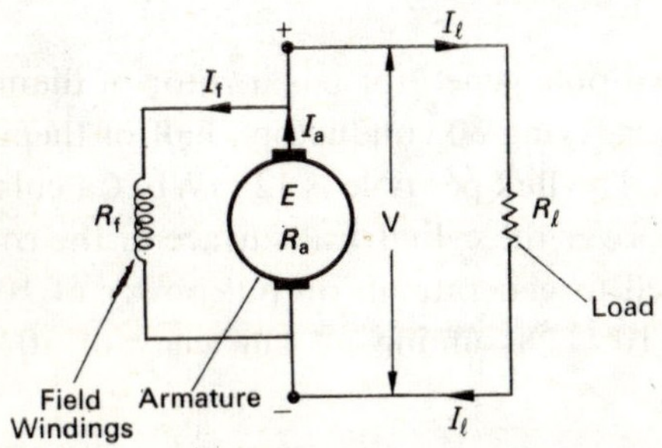

FIG. 8.5. D.C. shunt generator (ex. 8.13).

three parallel paths of the armature, field windings, and load, the common terminal p.d.

$$V = E - I_a R_a = I_f R_f = I_l R_l. \tag{8.5}$$

The mechanical power supplied in driving the rotor is converted into useful electrical output or load power VI_l, together with heat losses in various parts of the generator. These losses (in watts) may be listed as follows:

Air resistance, friction and heating of iron $= H$
Heating of armature conductors $= I_a^2 R_a$
Heating of field windings $= I_f^2 R_f$.

Then the total losses $= H + I_a^2 R_a + I_f^2 R_f$.

Solution

(a) From eqn. (8.5) the field current $I_f = \dfrac{V}{R_f} = \dfrac{110}{50} = 2.2$ A.

$\therefore$ field losses $I_f^2 R_f = 2.2^2 \times 50 = 240$ W. (a)

(b) From eqn. (8.4) the armature current $I_a = I_f + I_l$, where I_f is 2.2 A as in (a), and I_l is given as 20 A.

$$\therefore I_a = 2.2 + 20 = 22.2 \text{ A}$$

and the armature losses $I_a^2 R_a = 22.2^2 \times 0.05 = 25$ W. (b)

(c) Now the useful output power $V_l I_l = 110 \times 20 = 2200$ W and the efficiency is 81% or 0.81.

$$\therefore \text{ input power} = \frac{\text{useful output}}{\text{efficiency}} = \frac{2200}{0.81}$$

$$= 2700 \text{ W.} \qquad \text{(c)}$$

(d) Since the total losses $=$ input $-$ useful output

$= 2700 - 2200 = 500$ W

and "copper" losses $= (a) + (b) = 240 + 25 = 265$ W,

all other losses

$H = 500 - 265 = 240$ W (2 sig. fig.). (d)

8.14. A shunt generator delivers a power of 15 kW at a terminal p.d. of 200 V. The armature and field resistances are 0.05 Ω and 40 Ω respectively. Calculate the e.m.f. and current in the armature. [204 V; 80 A.]

8.15. The e.m.f. generated in a shunt generator is 500.0 V when supplying a current of 125 A. The field and armature resistances are 66 Ω and 0.040 Ω respectively. Calculate the terminal p.d. and field current. [494.7 V; 7.5 A.]

8.16. The no-load p.d. of a shunt generator at a certain speed is 117 V. When delivering a current of 145 A, at the same speed, the p.d. falls to 110 V. The field resistance is 22 Ω. Calculate the armature resistance. [0.048 Ω.]

8.17. The output of a small d.c. shunt generator is 10 A, 24 V. The armature and field resistances are 0.020 Ω and 20 Ω respectively. Calculate the friction and iron losses if the overall efficiency is 75%. [49 W.]

8.18. A shunt generator, having armature and field resistances of 0.10 Ω and 50 Ω respectively, delivers a load current of 86 A at a p.d. of 250 V. The torque required to drive the generator at the necessary speed of 1000 rev min^{-1} is 240 Nm. Calculate (a) the efficiency, and (b) the friction and iron losses. [(a) 86%; (b) 1500 W.]

CHAPTER 9

D.C. MOTORS

Torque

9.1. Worked Example

A two-pole d.c. motor has a useful flux per pole of 8.0 mWb. The rotor, which has a diameter of 10 cm and a length of 12 cm, has 40 armature conductors connected in series in each of two parallel paths. Calculate the mean torque when the armature current is 5.0 A.

Introduction and Solution

When a single conductor of length l lies perpendicularly across a magnetic field of flux density B and carries a current I, it experiences a force $F = BIl$ in a direction perpendicular to l and B. As in ex. 1.4, F is in newtons when B is in teslas, I is in amperes, and l is in metres.

In a motor with conductors distributed in slots in the armature, the flux density is greater at the teeth than in the slots, and also is greater at the pole faces than in the gap between each pole and the next. But if the armature makes one revolution, each conductor moves through all the flux φ which enters and leaves all teeth and slots. Therefore the mean flux density is used to obtain the mean force and torque.

$$\text{Thus mean flux density } B = \frac{\text{flux of one pole}}{\text{area of half rotor}} = \frac{\varphi}{\frac{1}{2}\pi dl},$$

where d = diameter and l = length of rotor.

$$\therefore\ B = \frac{8\times 10^{-3}}{\frac{1}{2}(\pi\times 0.1)\times 0.12} = 0.42\ \text{T}.$$

Now the commutator directs half the total current into each of the parallel paths through the conductors and ensures that each conductor has its current reversed as it passes from one pole face to the other. (Compare with the generator of ex. 8.2.) The forces on the two sides of the armature are then in the same direction of rotation, and all conductors contribute to the torque.

So, for each conductor,

$$\begin{aligned} \text{torque} &= \text{force} \times \text{radius} \\ &= BIl\times r \qquad (9.1) \\ &= (0.42\times \tfrac{5}{2}\times 0.12)\times 0.05 \text{ newton metre (Nm)} \\ &= 6.3\times 10^{-3}\ \text{Nm}. \end{aligned}$$

Then, for 80 conductors,

$$\begin{aligned} \text{the resultant mean torque} &= 80\times 6.3\times 10^{-3} \\ &= 0.50\ \text{Nm}. \end{aligned}$$

9.2. The rotor of a simple d.c. motor has two conductors each 10 cm long at a radius of 4.0 cm from the axis of rotation. Calculate the current for a maximum torque of 10^{-2} Nm in a field of flux density 0.80 T. [1.6 A.]

9.3. Calculate the force on a conductor of length 30 cm carrying a current of 4.0 A in a field of flux density 0.80 T when the length of the conductor is (a) perpendicular to the flux, and (b) at an angle of 60° to the direction of the flux. [(a) 0.96 N; (b) 0.83 N.]

9.4. Calculate the magnetic flux density required to produce a torque of 2.5 Nm with an armature of diameter 10 cm having 50 conductors each 10 cm long and carrying a current of 15 A. [0.67 T.]

Back E.M.F. and Speed

9.5. Worked Example

A 200 V d.c. shunt motor, having an armature resistance of 1.0 Ω, draws an armature current of 20 A when running at a speed of 1800 rev min^{-1}. Calculate the extra resistance required in series with the armature to reduce the speed to 1200 rev min^{-1} and yet maintain the same torque.

Introduction

When the armature of a machine rotates in the field of the stator, an e.m.f. is generated in the armature conductors whether the machine is in use as a generator or as a motor. In the case of a motor, this back e.m.f. E opposes the current I_a supplied to the armature. Then

$$V-E = I_a R_a, \tag{9.2}$$

for a shunt motor, where V is the p.d. applied at the terminals and R_a is the internal resistance of the armature.

Now, as in the generator of ex. 8.2, the e.m.f. E is determined by the rate at which flux is cut by the armature conductors.

Thus

$$E = \frac{2\varphi}{t} \tag{9.3}$$

for a two-pole machine in which φ is the flux in webers per pole and t is the time in seconds for one complete rotation past both poles.

Solution

At the speed of 1800 rev min^{-1} and current of 20 A, eqn. (9.2) gives

$$E = V-I_a R_a = 200-(20\times 1.0) = 180 \text{ V}.$$

Equation (9.3) shows that E is proportional to the rate of rotation. Therefore at the speed of 1200 rev min^{-1},

$$E = \frac{1200}{1800} \times 180 = 120\text{ V}.$$

Now with extra resistance R in series with R_a, eqn. (9.2) becomes

$$V - E = I_a(R_a + R). \tag{9.4}$$

But the requirement of constant torque implies a constant value of I_a (see ex. 9.1).

So, substituting in eqn. (9.4),

$$200 - 120 = 20(1.0 + R).$$

$$\therefore \; R = \frac{80}{20} - 1.0 = 3.0\,\Omega.$$

9.6. Worked Example

A 100 V d.c. series motor with total resistance of 0.5 Ω takes a current of 10 A at a speed of 2000 rev min^{-1}. Calculate the speed at which the torque drops to a quarter, assuming the flux to be proportional to the current.

Introduction

In a series motor the field windings are connected in series with the armature. Hence the flux φ and flux density B vary with the armature current I_a. In this example they are assumed to be proportional, i.e. $\varphi \propto I_a$. But the torque varies with both φ and I_a since the force F on each armature conductor is given by $F = BI_a l$, where B is the mean flux density and l is the length of each conductor.

$$\therefore \text{ torque} \propto BI_a \quad \text{and} \quad \varphi \propto B \propto I_a.$$

$$\therefore \text{ torque} \propto I_a^2.$$

Thus, when the torque drops to a quarter of the initial value, both I_a and φ drop to a half of their initial values.

Solution

From eqn. (9.4), when $I_a = 10$ A,

$$E = V - I_a(R_a + R) = 100 - (10 \times 0.5) = 95 \text{ V}$$

and at the new speed with half the current,

$$E = 100 - (5 \times 0.5) = 97.5 \text{ V}.$$

Further, from eqn. (9.3), $E \propto \varphi / T$, where t is the time of one revolution.

$$\therefore \text{ the rate of rotation } \frac{1}{t} \propto \frac{E}{\varphi}.$$

Thus, when E rises from 95 V to 97.5 V and φ is halved, the rate of rotation rises from 2000 rev min^{-1} to $2000 \times 97.5/95 \times 1/\frac{1}{2} =$ 4100 rev min^{-1}.

9.7. A 12 V d.c. shunt motor has an armature resistance of 0.50 Ω. When running with no load, the back e.m.f. is 11 V. Calculate the armature current (a) at the no-load speed, and (b) at half of the no-load speed. [(a) 2 A; (b) 13 A.]

9.8. The armature and field resistances of a 200 V d.c. shunt motor are 0.20 Ω and 200 Ω respectively. On no load, the total current is 7.5 A and the speed is 400 rev min^{-1}. Calculate the speed at which the current would be 100 A. [370 rev min^{-1}.]

9.9. A d.c. motor has an armature of resistance 0.20 Ω in parallel with field windings of resistance 100 Ω. A total current of 10 A is drawn at a p.d. of 250 V. Calculate the percentage change of current when the speed drops by 5%. [+630%.]

9.10. A 400 V d.c. series motor, having a total resistance of 0.50 Ω, runs at a speed of 100 rev min^{-1} when drawing a current of 40 A. Calculate the speed at which the current is 100 A, assuming that the flux per pole is 20 mWb when the current is 40 A and 30 mWb when 100 A. [610 rev min^{-1}.]

9.11. A shunt-wound d.c. motor for 100 V supply has an armature resistance of 0.20 Ω and a field resistance of 500 Ω. At a speed of 1200 rev min^{-1} the armature current is 20 A. (a) Calculate the extra resistance required in series with the field winding to increase the speed to 1800 rev min^{-1} for the same armature current. Assume that the flux is proportional to the field current. (b) Calculate the speed at which the original torque would be produced if extra resistance of 0.40 Ω were connected in series with the armature, and no extra in the field circuit. [(a) 250 Ω; (b) 1100 rev min^{-1}.]

Power and Efficiency

9.12. Worked Example

A 24 V d.c. shunt motor has armature resistance of 0.10 Ω, field resistance of 8 Ω, and draws a total current of 18 A. Calculate (a) the back e.m.f., (b) the mechanical power produced, and (c) the heat loss of the armature conductors.

Introduction

When the armature conductors of a motor sweep across the magnetic field, the generated e.m.f. E opposes the supply p.d. V, as shown by eqn. (9.2):

$$V - E = I_a R_a, \tag{9.5}$$

where I_a is the armature current and R_a is the resistance of the armature circuit. The electrical power required to drive the current I_a against the e.m.f. E is converted by the motor to mechanical power in rotating the armature. Thus in a shunt motor the mechanical power produced

$$EI_a = VI_a - I_a^2 R_a. \tag{9.6}$$

Here, VI_a is the total power supplied to the armature and $I_a^2 R_a$ is the power wasted in heating the armature conductors. Other losses are considered in ex. 9.13.

Solution

In the particular shunt motor, the supply current of 18 A divides into

$$\text{field current } I_f = \frac{V}{R_f} = \frac{24}{8} = 3\text{ A}$$

and

$$\text{armature current } I = 18-3 = 15\text{ A.}$$

(a) Then, from eqn. (9.5),

$$\text{back e.m.f. } E = V - I_a R_a$$
$$= 24-(15\times 0.10) = 22.5 = 23\text{ V} \quad \text{(2 sig. fig.).} \qquad \text{(a)}$$

(b) From eqn. (9.6),

$$\text{mechanical power produced} = EI_a$$
$$= 22.5\times 15 = 337.5 = 340\text{ W} \quad \text{(2 sig. fig.).} \qquad \text{(b)}$$

(c) As eqn. (9.6), the heat loss of the armature conductors is

$$I_a^2 R_a = 15^2\times 0.10 = 22.5 = 23\text{ W} \quad \text{(2 sig. fig.).} \qquad \text{(c)}$$

As a check, note that (b)+(c) = 337.5+22.5 = 360 W and total power supplied to armature = $VI_a = 24\times 15 = 360$ W.

9.13. Worked Example

A 200 V d.c. series motor has an armature resistance of 0.20 Ω and field resistance of 0.30 Ω. At a speed of 750 rev min^{-1} the current is 25 A and the friction and iron losses are 500 W. Calculate (a) the total copper loss, (b) the mechanical power output, (c) the output torque, and (d) the efficiency.

Introduction

The mechanical power developed in the armature of a motor is not entirely available as output power from the shaft. Some power loss occurs due to friction (and air resistance), and some due to the regeneration of electrical power in eddy currents and hysteresis effects in the magnetic material. This "iron loss" appears as heat, and is commonly considered with the heat due to friction, since both are largely independent of armature current although dependent on speed.

Thus available output power = mechanical power produced − friction and iron losses

$$= EI_a - H, \tag{9.7}$$

where E is the back e.m.f. in volts, I_a is the armature current in amperes, and H is the heating in watts due to friction and iron loss.

Solution

(a) The total "copper loss" due to armature and field resistances, R_a and R_f, in the series motor is $I_a^2(R_a+R_f)$ since the same current flows in both parts in series.

$$\therefore \text{ copper loss} = 25^2(0.20+0.30) = 310 \text{ W.} \tag{a}$$

(b) The mechanical power produced in the series motor [compare with eqn. (9.6)] is

$$\begin{aligned} EI_a &= VI_a - I_a^2(R_a+R_f) \\ &= (200\times 25) - 310 = 4700 \text{ W.} \end{aligned}$$

Then, from eqn. (9.7), the available output power

$$\begin{aligned} &= \text{mechanical power produced} - H \\ &= 4700 - 500 = 4200 \text{ W.} \end{aligned} \tag{b}$$

(c) Now the torque = P/ω, where P is the power (4200 W), and ω is the angular velocity corresponding to a speed of 750 rev min^{-1}, i.e. $\omega = 750\times 2\pi/60 = 78.6$ rad s^{-1}. Then

$$\text{torque} = \frac{4200}{78.6} = 53 \text{ Nm.} \tag{c}$$

$$\text{(d) Efficiency} = \frac{\text{useful output}}{\text{total input}}$$

$$= \frac{\text{mechanical power (b)}}{\text{(b)} + \text{copper loss (a)} + \text{friction and iron loss } (H)}$$

$$= \frac{4200}{4200+310+500}$$

$$= \frac{4.2}{5.01} = 0.84 \quad \text{or} \quad 84\%. \qquad \text{(d)}$$

9.14. A 400 V d.c. shunt motor having an armature resistance of 0.25 Ω, draws an armature current of 25 A at a speed of 600 rev min^{-1}. Calculate the torque produced, neglecting iron and friction losses. [160 Nm.]

9.15. A 1.5 V d.c. model motor running at a speed of 3000 rev min^{-1} draws a current of 0.60 A and gives an output power of 0.30 W. The field is provided by permanent magnets. Calculate (a) the efficiency, and (b) the torque in gram force-centimetre. Neglect friction and iron losses. [(a) 33%; (b) 9.7 gf. cm.]

9.16. A simple d.c. motor has a single rectangular coil of 30 turns, 6.0×4.0 cm, pivoted through the centres of the 4 cm sides, and turning in a field of flux density 0.60 T. The coil circuit is of 6.0 Ω resistance and the supply is of e.m.f. 18 V. Calculate (a) the maximum torque produced, and (b) the maximum mechanical power produced when running at a speed of 3000 rev min^{-1}. [(a) 0.13 Nm; (b) 10 W.]

9.17. A d.c. shunt motor having a field resistance of 200 Ω and an armature resistance of 2.5 Ω, takes a full load current of 2.6 A from 120 V supply. Assuming iron and friction losses to be negligible, calculate (a) the mechanical power produced at full load, (b) the efficiency at full load, and (c) the current if the motor were stalled. [(a) 230 W; (b) 74%; (c) 49 A.]

9.18. A 6.0 V d.c. shunt motor has armature resistance of 0.50 Ω and takes an armature current of 2.0 A when running on load at a certain speed. Calculate the decrease of mechanical power produced and the increase of speed when a change of load causes a 50% decrease of armature current. [−45%; +10%.]

9.19. A d.c. shunt motor having an armature resistance of 0.5 Ω runs on a supply of 200 V and develops an output power of 15 kW. Calculate the armature current assuming negligible friction and iron losses. [300 A or 100 A.]

9.20. A 200 V d.c. shunt motor has an armature resistance of 0.20 Ω and field resistance of 100 Ω. Friction and iron losses are 400 W when running with a supply current of 25 A. Calculate the output power and efficiency. [4.1 kW; 82%.]

9.21. The motor of the last exercise (ex. 9.20) is rewound with field coils of total resistance 0.20 Ω, which are connected in series with the armature of 0.20 Ω also. Calculate the output power and efficiency when the supply current is again 25 A at 200 V, with friction and iron losses of 400 W. [4.4 kW; 87%.]

CHAPTER 10

MAGNETIC FIELDS AND CIRCUITS

Straight Conductors

10.1. Worked Example

Two long parallel wires, 4.0 cm apart in air, carry currents of 5.0 A and 10.0 A respectively in the same direction. Calculate the resultant magnetizing force and flux density at a point 4.0 cm from each wire. Neglect other magnetic fields such as that of the Earth.

Introduction

The resultant field is formed of two components, one due to each current.

First, considering only one current, four important features of the magnetic field are detailed: (1) the flux density B, (2) the magnetomotive force (m.m.f.), (3) the magnetizing force H, and (4) the permeability μ.

(1) Rearrangement of eqn. (1.6) gives $B_1 = F_2/l_2 I_2$. Hence the flux density B in teslas (T) or webers per metre² (Wb m^{-2}) may be defined by the force per unit length F/l in N m^{-1} exerted by the flux on a conductor carrying current $I = 1$ A.

For the field at a point, due to current in a long thin straight conductor in vacuum (or air), eqn. (1.5) gives

$$B = 2\times 10^{-7}\frac{I}{r}, \tag{10.1}$$

where I is the current in the conductor in amperes and r is the radial distance in metres from the conductor to the point concerned.

(2) The current I is said to produce the flux by exerting a m.m.f. along each line of the flux. The magnitude of the resulting B depends also on the dimensions and properties of the medium. The magnitude of the m.m.f. is stated most conveniently by the current I in amperes. Thus

$$\text{m.m.f.} = I. \tag{10.2}$$

Important m.m.f.s are often those of coils, where the same current in numerous turns contributes numerous components to the resultant m.m.f. Hence the unit of m.m.f. is usually the "ampere-turn" (At).

(3) Around a long straight conductor in vacuum or air, the lines of flux are circular, shown in perspective in Fig. 1.3 (p. 5). The same m.m.f. is distributed along each line and is less effective the longer the line. This is stated by the magnetizing force

$$H = \frac{\text{m.m.f.}}{\text{flux path length}} = \frac{I}{2\pi r}, \tag{10.3}$$

where H is in ampere-turns per metre (At m^{-1}).

(4) Comparison of (10.1) and (10.3), shows $B \propto H$ in vacuum with a constant of proportionality called the permeability of free space μ_0. Thus, in free space,

$$\frac{B}{H} = \mu_0 = 4\pi \times 10^{-7}. \tag{10.4}$$

This value applies also to air with an error of less than 1 in 10^6.

Solution

The resultant field is found by the vector addition of the components of H or B due to the individual currents. Thus one component of the magnetizing force at P in Fig. 10.1 is due to the current of 5 A, and is given by eqn. (10.3):

$$H_5 = \frac{I}{2\pi r} = \frac{5}{2\pi \times 0.04}$$
$$= 20 \text{ At m}^{-1}.$$

Similarly, due to the other current, $H_{10} = 40$ At m^{-1}. The resultant of H_5 and H_{10} is given by the diagonal of the parallelogram of vectors. The components are tangential to the circular arcs centred

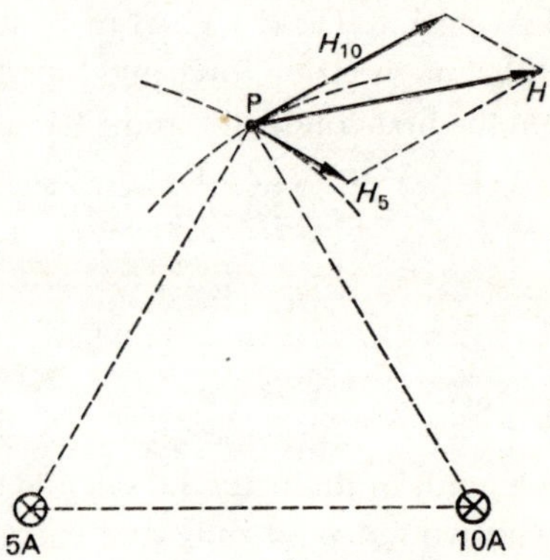

FIG. 10.1. Ex. 10.1.

on the wires and crossing at P at an angle of 60°. If the currents are perpendicular to and into the paper, both H_5 and H_{10} are in the plane of the paper and clockwise about their respective currents.

Then, from the cosine rule, the resultant magnetizing force H is given by

$$H^2 = 20^2 + 40^2 + (2 \times 20 \times 40 \cos 60°)$$
$$= 400 + 1600 + (1600 \times 0.5)$$
$$= 2800. \quad \therefore \; H = 53 \text{ At m}^{-1}.$$

The resultant flux density B is given by eqn. (10.4):

$$B = \mu_0 H = 4\pi \times 10^{-7} \times 53 = 6.7 \times 10^{-5} \text{ T}.$$

10.2. Two long, straight, parallel wires, separated by 2.0 cm in air, each carry a current of 10 A in the same direction. Calculate (a) the strength and direction of the force on a length of 12 cm of either wire, and (b) the magnetizing force of the field at either due to the other. [(a) 1.2×10^{-4} N; (b) 80 At m^{-1}.]

10.3. Two long, straight, parallel wires, carrying currents of 4.0 A and 6.0 A, are 10 cm apart in air. Calculate the magnetizing force and flux density at a point halfway between the wires when the currents are (a) in the same direction, and (b) in opposite directions. [(a) 6.4 At m^{-1}, 8 μT; (b) 32 At m^{-1}, 40 μT.]

10.4. Calculate the magnetizing force and flux density at a point situated 6.0 cm from the first conductor and 8.0 cm from the second conductor of ex. 10.3. [18 At m^{-1}, 22 μT.]

Air-cored Coils

10.5. Worked Example

An air-cored solenoid, of diameter 4.0 cm and of length 20 cm, has 300 turns of wire wound uniformly and carrying a current of 5.0 A. Calculate the magnetizing force (a) at one end of the solenoid, and (b) at the centre of the solenoid.

Introduction

The magnetizing force of a coil is obtained from Laplace's expression for that of a current element:

$$\delta H = \frac{I\delta l \sin \theta}{4\pi r^2}, \tag{10.5}$$

where δH is the magnetizing force produced at a point P distant r from a very small length δl of a thin conductor carrying a current I in a direction making an angle θ with the direction of r. For the direction of current shown in the plane of Fig. 10.2, the direction of δH is perpendicular to and into the paper at P.

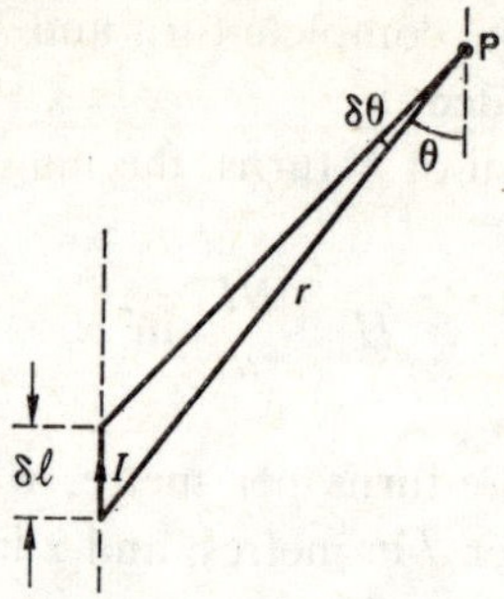

FIG. 10.2. Relating to a current element.

If δl were a part of a long, straight, thin conductor, it could be shown by integration that expression (10.5) leads to the previous expression (10.3).

For a single circular turn of wire, as in Fig. 10.3, H at a point X on the wire is found by summing the axial components $\delta H \sin \alpha$ due to all elements of the wire. The resultant of all radial components, $\delta H \cos \alpha$, is zero.

Thus
$$H = \Sigma \delta H \sin \alpha = \Sigma \left(\frac{I \delta l \sin \theta}{4 \pi r^2} \right) \sin \alpha.$$

But each element δl is perpendicular to the corresponding direction of r. So $\theta = 90°$ and $\sin \theta = 1$. Also $r = d/2 \sin \alpha$. So
$$H = \Sigma \frac{I \delta l}{\pi d^2} \sin^3 \alpha.$$

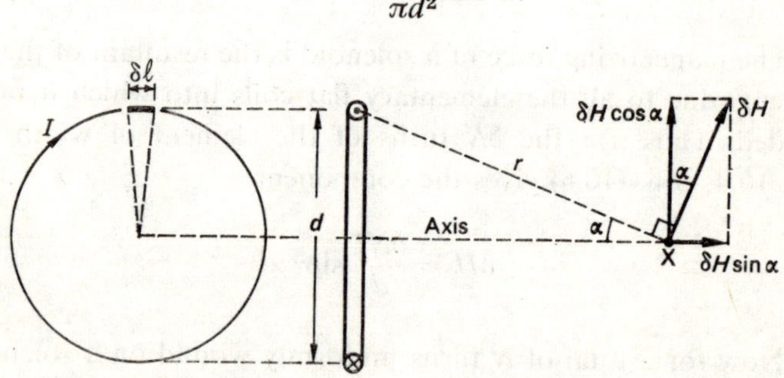

FIG. 10.3. Relating to a flat circular coil.

Now $\Sigma \delta l = \pi d$ for one complete turn and $N\pi d$ for N turns if these are practically coincident.

Thus for a flat coil of N turns, the magnetizing force along the axis at a point X is

$$H = \frac{NI}{d} \sin^3 \alpha, \tag{10.6}$$

where H is in ampere-turns per metre, when the current I is in amperes, the diameter d in metres, and α is the angle subtended at X by the radius of the coil.

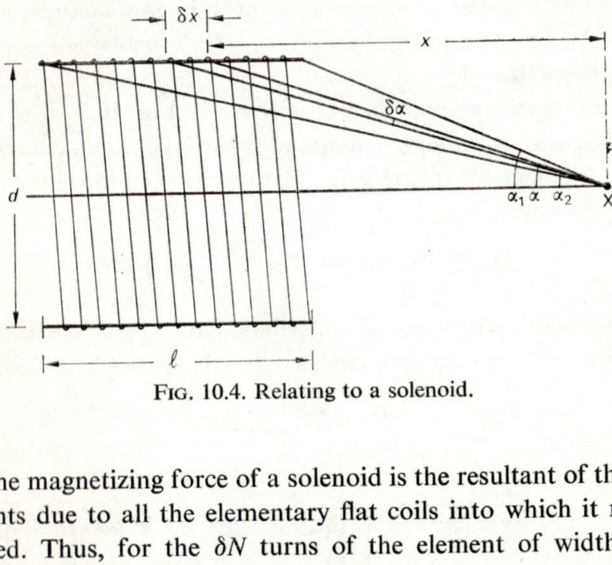

FIG. 10.4. Relating to a solenoid.

The magnetizing force of a solenoid is the resultant of the components due to all the elementary flat coils into which it may be divided. Thus, for the δN turns of the element of width δx in Fig. 10.4, eqn. (10.6) gives the component

$$\delta H = \frac{\delta N I}{d} \sin^3 \alpha.$$

Now for a total of N turns uniformly wound on a solenoid of length l, $\delta N = \delta x(N/l)$.

Also from Fig. 10.4, $\cot\alpha = \dfrac{x}{d/2}$ $\quad\therefore -\dfrac{1}{\sin^2\alpha}\,d\alpha = \dfrac{2}{d}\,dx.$

Then the resultant

$$H = \Sigma\,\delta H = \Sigma\left(\delta x \frac{N}{l}\,\frac{I}{d}\right)\sin^3\alpha$$

$$= \int_{\alpha_1}^{\alpha_2} \frac{N}{l}\,\frac{I}{d}\sin^3\alpha\left(-\frac{d}{2}\,\frac{1}{\sin^2\alpha}\,d\alpha\right) = \frac{NI}{2l}\int_{\alpha_1}^{\alpha_2} -\sin\alpha\,d\alpha$$

$$= \frac{NI}{2l}(\cos\alpha_2 - \cos\alpha_1), \tag{10.7}$$

where α_1 and α_2 are the angles subtended at X by the radii at the ends of the solenoid.

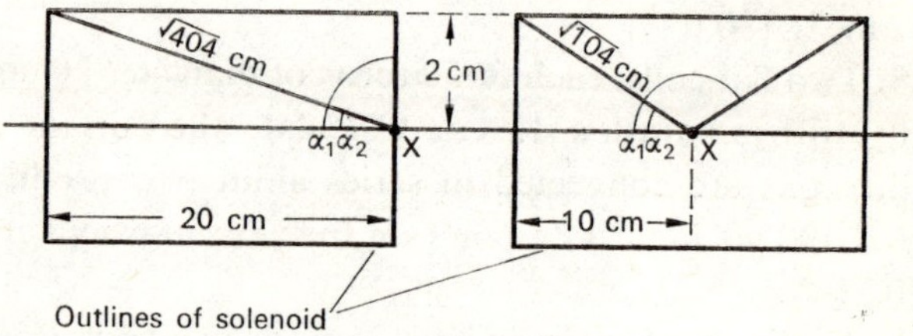

FIG. 10.5. Ex. 10.5.

At the centre of a solenoid for which $l \gg d$, α_2 would be approximately zero and α_1 approximately 180°, so giving

$$H = (NI)/l = \text{ampere turns per unit length.} \tag{10.8}$$

Solution

(a) Figure 10.5 shows that $\alpha_1 = 90°$ when X is at one end of the solenoid and $\cos\alpha_2 = 20/\sqrt{404} = 0.995$.

Then, from eqn. (10.7),

$$H = \frac{300\times 5}{2\times 0.20}(0.995-0) = 3700 \text{ At m}^{-1}. \tag{a}$$

(b) Figure 10.5 also shows that α_1 becomes obtuse when X is moved into the solenoid. When X is at the centre, $\cos\alpha_1 = -\cos\alpha_2$ and $\cos\alpha_2 = 10/\sqrt{104} = 0.98$.

Then, from eqn. (10.7),

$$H = \frac{300\times5}{2\times0.20}(0.98+0.98) = 7350 = 7400 \text{ At m}^{-1} \text{ (2 sig. fig.)}$$

(b)

Note that 7350 is 2% less than would be obtained by the approximation of eqn. (10.8), i.e. $H \triangleq 7500$ At m^{-1}.

10.6. Calculate the current required in a flat coil of 25 turns of diameter 10 cm to produce a flux density of 1.0 mT in air at its centre. [3.2 A.]

10.7. A flat coil has 20 turns of diameter 15 cm and carries a current of 5.0 A. Calculate the magnetizing force and flux density in air at a point on the axis at a distance of 10 cm from the centre. [140 At m^{-1}; 180 μT.]

10.8. Two flat coils, each of 50 turns of diameter 20 cm, are set coaxially with separation 15 cm. Calculate the current required, when the coils are connected in series aiding, to produce a flux density of 100 μT in air at a point on the axis midway between the coils. [0.31 A.]

10.9. An air-cored solenoid has 125 turns wound uniformly on a non-magnetic tube of length 25 cm and of diameter 8.0 cm. Calculate the current required to produce a magnetizing force of 1000 At m^{-1} at the centre. [2.1 A.]

10.10. A solenoid has a diameter of 10 cm and a length of 20 cm. Calculate the magnetizing force at a point on the axis, 5.0 cm from the centre, as a percentage of that at the centre. [93%.]

Permeability and Reluctance

10.11. Worked Example

A steel ring, having a mean diameter of 10 cm and a cross-section area of 4.0 cm², is wound uniformly with 200 turns of wire. The relative permeability of the steel is 1250. Calculate (a) the current required to produce a flux of 0.60 mWb in the ring, and (b) the reluctance of the ring.

Introduction and Solution

A ring winding, as in Fig. 10.6, may be imagined to be formed by bending a long, straight solenoid until the ends meet. Since the solenoid is then effectively endless, the magnetizing force H is

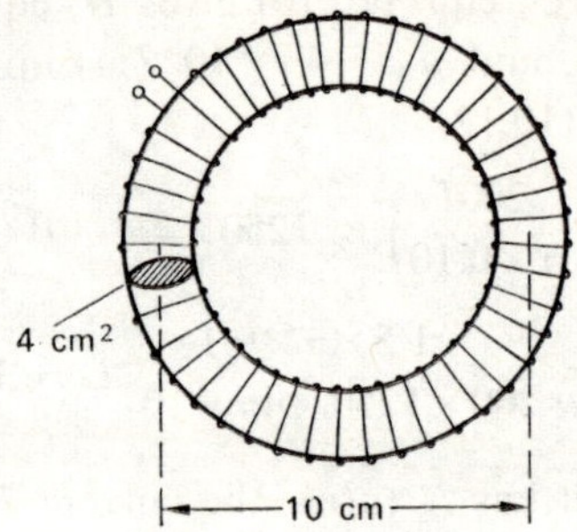

FIG. 10.6. Ex. 10.11.

uniform and equal to that at the centre of the equivalent infinitely long, straight solenoid. Thus, from eqn. (10.8),

$$H = \frac{NI}{l} = \frac{NI}{\pi d}, \tag{10.9}$$

where I is the current flowing in N turns and d is the mean diameter in metres. Thus, for the particular ring,

$$H = \frac{200I}{\pi \times 0.10} \text{ At m}^{-1}. \tag{10.10}$$

The density of the flux, which is almost entirely confined within the ring, is given by

$$B = \frac{\text{flux}}{\text{cross-section area}} = \frac{0.6 \times 10^{-3}}{4 \times 10^{-4}}$$

$$= 1.5 \text{ T}. \tag{10.11}$$

The ratio B/H is the permeability of the medium μ. This depends upon two factors: μ_0, determined by the units of B and H, and μ_r, determined by the nature of the particular medium. In vacuum only μ_0 applies, but, in general,

$$B/H = \mu = \mu_r \mu_0 , \tag{10.12}$$

where μ_r is called the relative permeability of the medium.

Thus, for the steel, eqn. (10.10) gives H, eqn. (10.11) gives B, μ_r is given as 1250, and $\mu_0 = 4\pi \times 10^{-7}$ as in eqn. (10.4). Then, substituting in eqn. (10.12),

$$1.5 \Big/ \left(\frac{200I}{\pi \times 0.10} \right) = 1250 \times 4\pi \times 10^{-7}$$

$$\therefore \; I = \frac{1.5 \times \pi \times 0.1}{200 \times 1250 \times 4\pi \times 10^{-7}} = 1.5 \text{ A}. \tag{a}$$

(b) Since the coil has N turns, the current I encircles the flux and the flux encircles the current N times. Thus the total current linkage $= NI$.

Ampere's theorem states that the magnetomotive force along any closed path equals the total current linkage for that path.

So, from eqn. (10.9),

$$\text{m.m.f.} = NI = Hl \tag{10.13}$$

for a closed path of constant H and length l.

Now, from eqn. (10.12), $H = B/\mu$. Also $B = \varphi/A$, where φ is the total flux through the cross-section area A.

Then, substituting in eqn. (10.13),

$$NI = \varphi \times \left(\frac{l}{\mu A} \right). \tag{10.14}$$

The last term in brackets depends on the magnetic material and its dimensions, and is called the reluctance, by analogy with the resistance of an electrical conductor for which $R = l(\sigma A)$ where σ is the conductivity.

Then eqn. (10.14) may be put in words:

$$\text{magnetomotive force} = \text{flux} \times \text{reluctance}. \qquad (10.15)$$

This is analogous to the electrical relationship:

$$\text{electromotive force} = \text{current} \times \text{resistance}.$$

So the reluctance of the ring is

$$\frac{l}{\mu A} = \frac{\pi \times 0.10}{(1250 \times 4\pi \times 10^{-7}) \times 4 \times 10^{-4}} = 5 \times 10^5 \text{ At Wb}^{-1}. \qquad \text{(b)}$$

The units of reluctance and the value just obtained may be checked by rearranging eqn. (10.15):

$$\text{reluctance} = \frac{\text{m.m.f. (in At)}}{\text{flux (in Wb)}} = \frac{200 \times 1.5}{0.6 \times 10^{-3}} = 5 \times 10^5 \text{ At Wb}^{-1}.$$

10.12. Calculate the flux density in a long steel bar of relative permeability 1800 when the magnetizing force is 500 At m^{-1}. [1.13 T.]

10.13. A toroidal winding on a steel ring of mean diameter 12 cm and cross-section area 5.0 cm^2 has 100 turns and carries a current of 5.0 A. Calculate (a) the m.m.f., (b) the reluctance, and (c) the flux. Assume a relative permeability of 750. [(a) 500 At; (b) 8×10^5 At Wb^{-1}; (c) 0.63 mWb.]

10.14. An air gap of width 1.0 mm has an effective cross-section area of 10 cm^2 and carries a flux of 2.5 mWb. Calculate (a) the reluctance, and (b) the m.m.f. required. [(a) 4.0×10^5 At Wb^{-1}; (b) 1000 At.]

10.15. A long silicon iron rod of diameter 1.0 cm lies in a long solenoid wound with 5 layers of wire each having 10 turns per cm. Calculate the current required to produce a flux density of 1.1 T at the centre assuming a relative permeability of 3500. When the current is 300 mA, the flux in the rod is found to be 0.11 mWb. Calculate the relative permeability in this condition. [50 mA; 740.]

10.16. A solenoid, 50 cm long, is wound uniformly with 2000 turns on a rod of cross-section area 1.0 cm². The magnetic characteristic of the material is given:

H (At m^{-1})	100	200	300	500	700	1000	1500	2000
B (T)	0.35	0.65	0.82	1.01	1.12	1.20	1.27	1.30

Determine the flux through the rod near its centre and the relative permeability when the current is 160 mA. [0.11 mWb; 1400.]

Series Circuits

10.17. Worked Example

A ring core with an air gap is wound with a coil of 300 turns. The mean length of the flux path in the core is 50 cm, and the cross-section area is 10 cm². The mean cross-section area of the flux in the gap is 20 cm² for a length of 6.0 mm. For the same material as ex. 10.16 and for a gap flux density of 0.50 T, calculate (a) the relative permeability of the core, (b) the reluctance of the core, and (c) the current required.

Introduction

In many magnetic circuits, having two or more parts in series, with different cross-sections, and/or different materials, approximately the same flux passes through each section. Some leakage flux may bypass a section by a "parallel" path in air, just as electric current may leak through an insulator in parallel with the main circuit of conductors in series. However, in the following problems leakage flux is to be neglected unless the contrary is stated.

The reluctance of each part of a series magnetic circuit may be reckoned to cause a magnetic potential drop, m.p.d., along its length, and so to require a part of the m.m.f. to maintain the flux. For each part the m.p.d. = flux × reluctance, just as electrical p.d.

= current×resistance. Thus eqn. (10.13) and (10.14) may be modified to apply to two parts in a series circuit:

$$(\text{m.p.d.})_1 = H_1 l_1 = \varphi \times [l_1/(\mu_1 A_1)], \tag{10.16}$$

$$(\text{m.p.d.})_2 = H_2 l_2 = \varphi \times [l_2/(\mu_2 A_2)], \tag{10.17}$$

where φ is the same in each equation.

The m.m.f. of the coil NI equals the total m.p.d. Thus, for a two-part circuit,

$$NI = H_1 l_1 + H_2 l_2 = \varphi(l_1/\mu_1 A_1 + l_2/\mu_2 A_2). \tag{10.18}$$

Note the similarity to Kirchhoff's equation for a series electric circuit (see ex. 3.27):

$$E = V_1 + V_2 = I(R_1 + R_2).$$

Note also that although m.p.d. and m.m.f. may often be reckoned separately, the m.m.f. is usually distributed continuously along the circuit. This is more like the induced e.m.f. in the long wire of a transformer winding than the e.m.f. at the plates of a battery.

Solution

(a) The air gap of the ring core has $B_1 = 0.5$ T and $A_1 = 20 \times 10^{-4}$ m².

$$\therefore \ \varphi = B_1 A_1 = 0.5 \times 20 \times 10^{-4} = 10^{-3} \text{ Wb}.$$

For the same flux in the metal core,

$$B_2 = \frac{\varphi}{A_2} = \frac{10^{-3}}{10 \times 10^{-4}} = 1 \text{ T}.$$

By plotting a graph of the magnetic characteristic from the tabulated figures of ex. 10.16, it is seen that $B_2 = 1$ T when $H_2 = 485$ At m^{-1}. Then the core has permeability $\mu_2 = B_2/H_2 = 1/485$, and relative permeability

$$\mu_r = \frac{\mu_2}{\mu_0} = \frac{1}{485 \times 4\pi \times 10^{-7}} = 1600. \tag{a}$$

(b) The core has reluctance, as in eqn. (10.17),

$$\frac{l_2}{\mu_2 A_2} = \frac{50\times10^{-2}}{\frac{1}{485}\times(10\times10^{-4})} = 2.4\times10^5 \text{ At Wb}^{-1}. \qquad \text{(b)}$$

(c) The air gap has reluctance, as in eqn. (10.16):

$$\frac{l_1}{\mu_1 A_1} = \frac{6\times10^{-3}}{(4\pi\times10^{-7})\times(20\times10^{-4})} = 2.39\times10^6 \text{ At Wb}^{-1}.$$

Then, using eqn. (10.18), $I = \frac{\varphi}{N}\left(\frac{l_1}{\mu_1 A_1}+\frac{l_2}{\mu_2 A_2}\right)$

$$= \frac{10^{-3}}{300}[(2.39\times10^6)+(2.4\times10^5)]$$

$$= \frac{2630}{300} = 8.8 \text{ A}. \qquad \text{(c)}$$

10.18. A coil of 1000 turns is wound on a steel ring of mean diameter 15 cm and cross-section area 5.0 cm². Calculate (a) the magnetic flux when the current is 0.15 A, and (b) the current required for the same flux when the ring has an air gap of 1.0 mm. Assume the relative permeability to be 500. [(a) 10^{-4} Wb; (b) 0.31 A.]

10.19. An electric relay has an iron circuit of approximately constant cross-section and relative permeability of 500. The flux path has a length of 15 cm in the iron and 1.0 mm across an air gap. Neglecting leakage, calculate the flux density in the gap when a current of 50 mA flows in the coil of 5000 turns. [0.24 T.]

10.20. A toroidal coil of length 1.0 m has 1000 turns and carries a current of 2.0 A. Calculate the flux density and magnetizing force within the coil when the core consists of (a) air, (b) iron of relative permeability 200, and (c) as (b) but with an air gap of 2.0 mm. [(a) 2.5×10^{-3} T, 2000 At m^{-1}; (b) 0.5 T, 2000 At m^{-1}; (c) 0.36 T, 1.4×10^3, 2.9×10^5 At m^{-1}.]

10.21. A rectangular transformer core has a flux path length of 60 cm and cross-section 3.0×4.0 cm. Two 400-turn coils are mounted on opposite sides of the core and carry a current of 60 mA in series aiding. The flux produced is 0.72 mWb. Two air gaps each of 0.25 mm are then introduced by opening the joints of the core laminations. Assuming negligible fringing and negligible change of permeability, calculate the final magnetizing force and flux density in the core. [13 At m^{-1}; 0.1 T.]

10.22. An electromagnet consists of two cylindrical steel-cored coils, each of 500 turns, mounted vertically side by side on a steel base, and each fitted with a pole piece at the upper end of the core. The dimensions are as follows: each core has length 10 cm and diameter 3.0 cm; the base has a flux path length of 10 cm with effective cross-section area of 5.0 cm^2; each pole has a flux path length of 4.0 cm and mean cross-section area of 6.0 cm^2; the air gap between the poles is 1.0 cm wide with effective cross-section area of 20 cm^2. Calculate the current required in the coils, which are connected in series aiding, for a mean flux density of 0.30 T in the gap. The magnetic characteristic is given in ex. 10.16. [2.6 A.]

Magnetic Load Line

10.23. Worked Example

The core of an inductor has a flux path length of 25 cm in magnetic material, in series with an air gap of 0.50 mm. Assuming the cross-section of the flux to be constant, determine the flux density produced by an m.m.f. of 500 At. The magnetic characteristic of the material is given by ex. 10.16.

Solution

The m.m.f. of the coil equals the sum of the m.p.d.s. of the air gap and the material, which carry the same flux density B in series. Thus eqn. (10.18) gives $Hl + H_0 l_0 = NI$. Here

the m.m.f. of the coil $= NI = 500$ At,

the m.p.d. for the material $= Hl = H \times 0.25$ At,

and

$$\text{the m.p.d. for the air gap} = H_0 l_0 = \frac{B}{\mu_0} l_0$$

$$= \frac{B}{4\pi \times 10^{-7}} \times 5 \times 10^{-4}.$$

$$\therefore\ (H \times 0.25) + \left(\frac{B \times 5 \times 10^{-4}}{4\pi \times 10^{-7}}\right) = 500.$$

$$\therefore\ H + 1592B = 2000. \qquad (10.19)$$

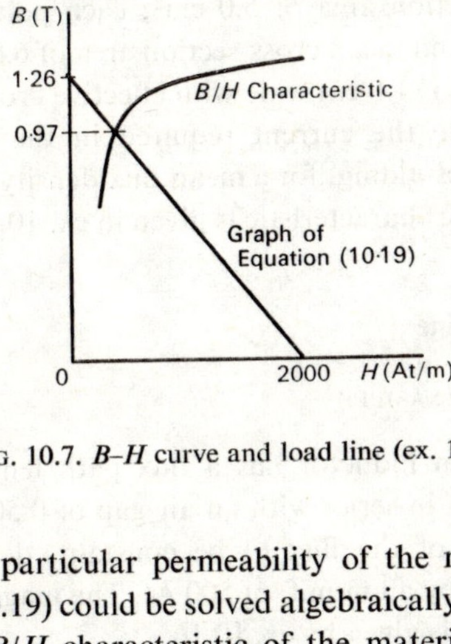

FIG. 10.7. *B–H* curve and load line (ex. 10.23).

Now if the particular permeability of the material B/H were known, eqn. (10.19) could be solved algebraically; but this is not so. However, the B/H characteristic of the material is given by the figures of ex. 10.16. A graphical solution is then obtained by finding the intersection of the B/H characteristic and the "load line" of eqn. (10.19) as shown in Fig. 10.7. The load line is drawn easily through the points where eqn. (10.19) has $B = 0$ and $H = 0$, i.e. $(B = 0,\ H = 2000)$ and $\left(H = 0,\ B = \frac{2000}{1592} = 1.26\right)$.

The intersection then gives the flux density of 0.97 T.

10.24. Determine the flux density in the relay of ex. 10.19, assuming the permeability to be unknown, and assuming the B/H characteristic of the material to be given by the figures of ex. 10.16. [0.3 T.]

Energy and Attraction

10.25. Worked Example

Two magnetic pole faces, each of area 5 cm², face each other across a 2 mm air gap which carries a flux of 0.25 mWb. Calculate (a) the energy stored in the gap, and (b) the force of attraction between the poles.

Introduction

Energy is required to establish a magnetic field in a coil since the magnetizing current must be built up against a back e.m.f. The required energy

$$W = \int_0^t ei\, dt,$$

[see eqn. (11.12)] where e and i are the values of e.m.f. and current at time t during the growth, and ei is the power at that instant. But $e = N(d\varphi/dt)$, where N is the number of turns of the coil and $d\varphi/dt$ is the rate of change of flux [see eqn. (11.1)].

$$\therefore\ W = \int_0^t Ni\frac{d\varphi}{dt}\, dt = \int_0^\varphi Ni\, d\varphi.$$

Further, if l and A are the length and cross-section area of the path of the flux, which is assumed to be uniform, $\dfrac{Ni}{l}$ = magnetizing force H, and $\dfrac{\varphi}{A}$ = flux density B.

$$\therefore\ W = \int_0^{BA} Hl\, d(BA).$$

But (lA) is the volume of the field. Therefore the energy per unit volume

$$\frac{W}{lA} = \int_0^B H\,dB. \tag{10.20}$$

This may be integrated, for a constant permeability, as in air, by putting $H = B/\mu$.

Thus the energy per unit volume in joules per cubic metre

$$= \int_0^B \frac{B}{\mu}\,dB = \frac{B^2}{2\mu} = \frac{1}{2}\,BH. \tag{10.21}$$

Note that when μ is not constant, as in a ferromagnetic material, the integral (10.20) may be found graphically from the B/H loop. In particular, for one cycle around the B/H loop, $\int H\,dB$ is the area enclosed by the loop, calculated to scale. Thus the area of the loop represents the energy per unit volume taken per cycle. This energy reappears as heat "hysteresis loss".

If the volume of the field between two poles is reduced by reducing the gap, without changing B or H, as between two rods inserted in opposite ends of a solenoid, stored energy is released from the volume swept out. Thus if the gap width is reduced by δl and B and H are maintained constant, the energy in joules released from the gap is

$$(\text{energy per unit volume})\times(\text{swept volume}) = (\tfrac{1}{2}BH)\times(A\,\delta l).$$

This energy is used in the work $F\,\delta l$ done by the force of attraction F as the poles move closer together.

$$\therefore\ \tfrac{1}{2}BHA\,\delta l = F\,\delta l.$$

$$\therefore\ \text{force per unit area} = F/A = \tfrac{1}{2}BH = \text{energy per unit volume}. \tag{10.22}$$

Note that the unit, newton per square metre, for the force per unit area, is equivalent to the unit, joule per cubic metre for the energy per unit volume.

Solution

(a) In the air gap, where $\varphi = 0.25\times10^{-3}$ Wb,

$$B = \frac{\varphi}{A} = \frac{0.25\times10^{-3}}{5\times10^{-4}} = 0.5 \text{ T}$$

and
$$H = \frac{B}{\mu_0} = \frac{0.5}{4\pi\times10^{-7}} \quad \text{At m}^{-1}.$$

$$\therefore\ \frac{1}{2}BH = \frac{1}{2}\times0.5\times\frac{0.5}{4\pi\times10^{-7}} = 10^5 \text{ J m}^{-3} \quad \text{or} \quad \text{N m}^{-2}.$$

The volume of the field $= Al = (5\times10^{-4})\times(2\times10^{-3})$
$= 10^{-6}$ m^3.

Therefore from eqn. (10.21), stored energy $= \left(\frac{1}{2}BH\right)\times(Al)$
$= 10^5\times10^{-6}$
$= 0.1$ J. (a)

(b) From eqn. (10.22) the force of attraction equals

$$\left(\tfrac{1}{2}BH\right)A = 10^5\times5\times10^{-4} = 50 \text{ N}. \qquad \text{(b)}$$

10.26. Two long iron rods, each of diameter 2.0 cm, are inserted axially through opposite ends of a long solenoid, wound with ten turns of wire per cm, and carrying a current of 2.0 A. Calculate the force of attraction of the rods when their ends are close to the centre of the solenoid. Assume a relative permeability of 600. [280 N.]

10.27. The core of a transformer for 50 Hz operation has a mass of 15 kg and a density of 7.8 g cm^{-3}. The hysteresis loop area is 20 cm^2, when plotted on a graph scaled as follows: for B, 1 cm $\equiv$ 0.2 T; for H, 1 cm $\equiv$ 200 At m^{-1}. Calculate the power loss due to hysteresis. [77 W.]

10.28. A small lifting magnet, having two poles each of area 50 cm^2, supports a steel plate of weight 50 kgf, with an air gap of 1.0 mm between each pole and the plate. Assuming the reluctance of the magnet and plate to be 10^5 At Wb^{-1}, calculate (a) the energy stored in the air gaps, and (b) the magnetomotive force to be supplied by the coil. [(a) 0.49 J; (b) 730 At.]

Parallel Circuits

10.29. Worked Example

A silicon–iron transformer core, having the dimensions shown in Fig. 10.8, has a coil of 1000 turns around the centre limb. Calculate the current required to produce a flux density of 1.4 T in an air gap of width 1.0 mm, when this gap is (a) at *X*, or (b) at *Y*. Neglect fringing and leakage. The magnetic characteristic of the silicon-iron is given by the following:

H (At m^{-1})	500	700	1000	1500	3000	4500	6500
B (T)	1.17	1.25	1.33	1.40	1.50	1.55	1.60

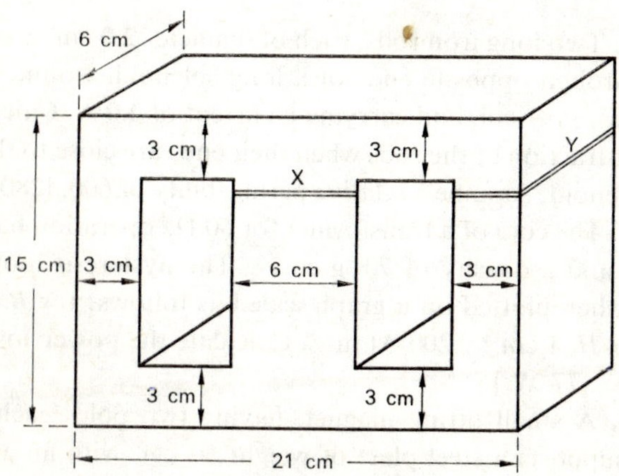

FIG. 10.8. Ex. 10.29.

Solution

Figure 10.9 shows the directions of the flux in the two parallel paths and the centre limb of the core with the estimated mean lengths of these paths found by considering the middle of the cross-section of each part. These lengths could be reduced to allow for the curvature of the flux at the corners, but the effect is assumed to be negligible.

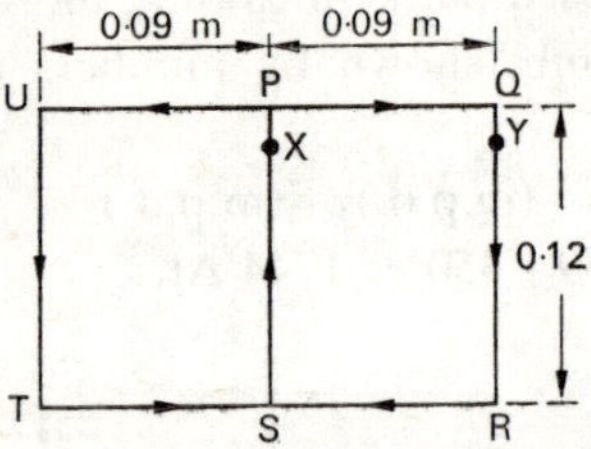

FIG. 10.9. Flux directions and mean path lengths (ex. 10.29).

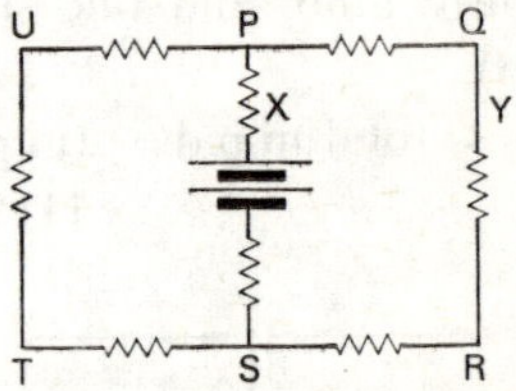

FIG. 10.10. Equivalent electric circuit of ex. 10.29.

Although the m.m.f. is distributed around the circuit it is convenient and more equivalent to imagine the m.m.f. to be concentrated near the centre of the centre limb, like the battery in the equivalent electrical circuit of Fig. 10.10.

(a) In the air gap at X,

$$H_0 = \frac{B}{\mu_0} = \frac{1.4}{4\pi \times 10^{-7}} \text{ At m}^{-1} \quad \text{and} \quad l_0 = 10^{-3} \text{ m}.$$

Therefore for the gap

$$(\text{m.p.d.})_0 = H_0 l_0 = \frac{1.4}{4\pi \times 10^{-7}} \times 10^{-3} = 1115 \text{ At}.$$

The centre limb carries the same flux density in series with the gap, and has $l_1 = 0.12 - 10^{-3}$ m, with H_1 obtained from the magnetic characteristic where $B = 1.4$ T, i.e. $H_1 = 1500$ At m^{-1}.

Therefore for the centre limb

$$(\text{m.p.d.})_1 = H_1 l_1 = 1500 \times 0.119 = 179 \text{ At}.$$

Now for the parallel paths *PQRS* and *PUTS* the m.p.d. is the same as the electrical p.d. would be in the analogous electric circuit of Fig. 10.10. This $(\text{m.p.d.})_2 = H_2 l_2$, where $H_2 = 1500$ At m^{-1} since B is the same throughout the circuit, and $l_2 = 0.09+0.12+0.09 = 0.30$ m.

Thus for either path *PQRS* or *PUTS*

$$(\text{m.p.d.})_2 = 1500\times 0.30 = 450 \text{ At.}$$

Then, for either of the two closed paths, including the air gap, the centre limb, and one side limb, shown by numbers in Fig. 10.11(a),

$$\text{total m.p.d.} = (\text{m.p.d.})_0+(\text{m.p.d.})_1+(\text{m.p.d.})_2$$
$$= 1115+179+450 = 1744 \text{ At.}$$

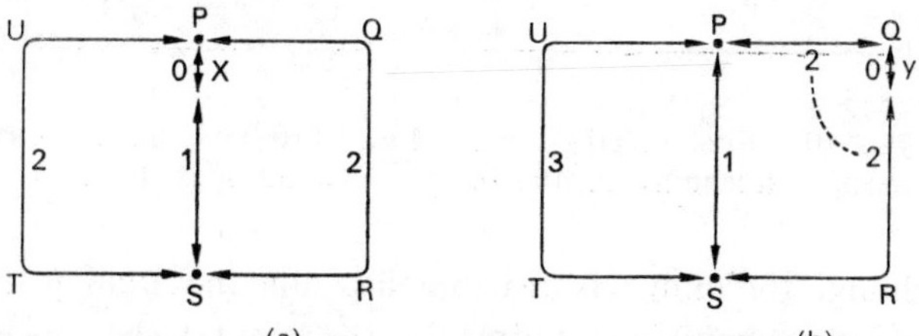

FIG. 10.11. (a) Ex. 10.29(a). (b) Ex. 10.29(b).

The coil must provide an m.m.f. of this value, and so a current I is required, where $NI = 1744$, and $N = 1000$ turns.

$$\therefore\ I = \frac{1744}{1000} = 1.7 \text{ A.} \qquad \text{(a)}$$

(b) When the air gap is at Y, carrying the same density of flux as at X, $(\text{m.p.d.})_0$ for the gap is again 1115 At. This is now in series with $(\text{m.p.d.})_2$ for flux of the same density in the solid material of the path *PQRS*, shown in Fig. 10.11(b). Now $(\text{m.p.d.})_2 = H_2 l_2$, where $H_2 = 1500$ At m^{-1} as before, but $l_2 = 0.30-10^{-3}$ m.

$$\therefore\ (\text{m.p.d.})_2 = 1500\times 0.299 = 449 \text{ At.}$$

Then, for the solid and the air in series, in the path *PQRS*,

$$(\text{m.p.d.})_3 = (\text{m.p.d.})_0+(\text{m.p.d.})_2 = 1115+449 = 1564 \text{ At.}$$

The same m.p.d. applies to the path *PUTS*, which therefore has a magnetizing force

$$H_3 = \frac{(\text{m.p.d.})_3}{\text{length}} = \frac{1564}{0.30} = 5220 \text{ At m}^{-1}.$$

The magnetic characteristic, obtained by plotting the given values of B and H, indicates $B_3 = 1.57$ T when $H_3 = 5220$ At m^{-1}. So the path *PUTS* has a higher value of B than the path *PQRS*, in the unsymmetrical case (b).

Now the total flux in the centre limb is the sum of the fluxes in the side limbs.

$$\therefore \; B_1A_1 = B_2A_2 + B_3A_3,$$

i.e. $$B_1 \times 36 \times 10^{-4} = (1.4 \times 18 \times 10^{-4}) + (1.57 \times 18 \times 10^{-4}).$$

$$\therefore \; B_1 = \frac{1.4+1.57}{2} = 1.485 \text{ T}.$$

Then $H_1 = 2700$ At m^{-1} from the B/H characteristic,

and $(\text{m.p.d.})_1 = H_1 l_1 = 2700 \times 0.12 = 324$ At.

Now the m.m.f. to be provided by the coil equals

$$(\text{m.p.d.})_3 + (\text{m.p.d.})_1 = 1564 + 324 = 1888 \text{ At}.$$

Therefore, in 1000 turns, the current must be $\dfrac{1888}{1000} = 1.9$ A. (b)

10.30. A coil of 2000 turns is mounted on a silicon–iron core which has a flux path of cross-section 3.0×4.0 cm. The path length is 40 cm in the core, and 2.0 cm in an air gap in series with the core. A magnetic shunt, having a reluctance of 10^7 At Wb^{-1}, is clamped in parallel with the air gap. Calculate the current required in the coil to produce a flux density of 0.50 T in the air gap, assuming the effective cross-section area of the gap to be 50% greater than that of the core. The magnetic characteristic is given in ex. 10.29. [4.3 A.]

10.31. A choke core, made of silicon–iron laminations, shown in Fig. 10.12, is built to a thickness of 2.0 cm. Calculate the current in a coil of 500 turns around the centre limb, for a flux of 0.60 mWb through limb *A*, with (a) a closed iron circuit, and (b) an air gap of 0.50 mm at *A*. The magnetic characteristic is given in ex. 10.29. [(a) 1.7 A; (b) 3.1 A.]

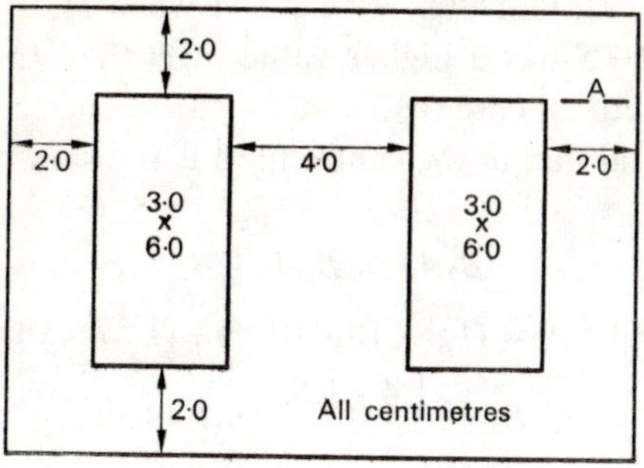

FIG. 10.12. Ex. 10.31.

Permanent Magnets

10.32. WORKED EXAMPLE

Calculate the length and cross-section area of a permanent magnet, to be made of material operating at the optimum values of remanent flux density and demagnetizing force, 1.0 T and 4.0×10^4 At m^{-1} respectively. The air gap of the magnet is to have a length of 1.0 cm and a cross-section area of 3.0 cm^2, and is to carry a flux of density 0.60 T.

Introduction

Since the total flux across the gap is supplied through the magnet material,

$$B_g A_g = B_m A_m, \tag{10.23}$$

where B_g and A_g are the flux density and cross-section area of the gap, and B_m and A_m are the corresponding values for the same total flux through the magnet.

The pole faces of the magnet have a magnetic potential difference which applies both forward across the gap and backward through the magnet. Looking forward, the m.p.d. $= H_g l_g$, where l_g is the length of the gap, and H_g is the magnetizing force in the gap, as in Fig. 10.13. Assuming that the pole pieces have negligible reluctance, and looking backward, the same m.p.d. $= H_m l_m$, where l_m is the length of the path through the magnet and H_m is the demagnetizing force of the free poles.

$$\therefore\ H_g l_g = H_m l_m. \tag{10.24}$$

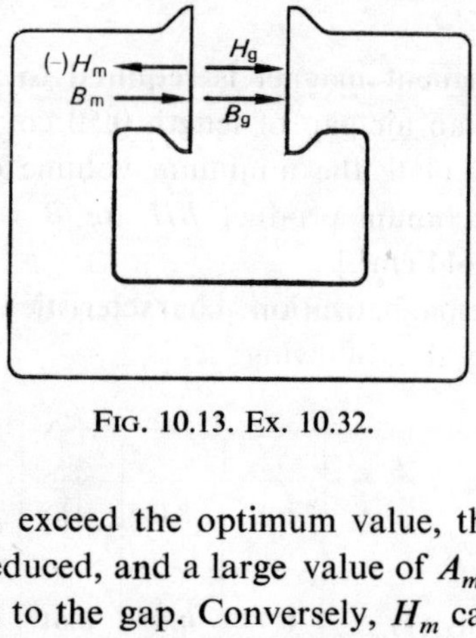

FIG. 10.13. Ex. 10.32.

If H_m were to exceed the optimum value, the available value of B_m would be reduced, and a large value of A_m would be needed to supply the flux to the gap. Conversely, H_m cannot be less than the optimum without the need to increase l_m.

Now from eqns. (10.23) and (10.24) the volume of magnet material

$$A_m l_m = \frac{B_g A_g}{B_m} \times \frac{H_g l_g}{H_m}.$$

Therefore the volume is a minimum when the product $(B_m H_m)$ is a maximum. The individual values of B_m and H_m for this condition are given in the question.

Solution

From eqn. (10.23),

$$A_m = \frac{B_g A_g}{B_m} = \frac{0.6\times(3\times10^{-4})}{1.0}$$

$$= 1.8\times10^{-4}\ \text{m}^2 = 1.8\ \text{cm}^2.$$

Also, from eqn. (10.24),

$$l_m = \frac{H_g l_g}{H_m} = \frac{[0.6/(4\pi\times10^{-7})]\times10^{-2}}{4\times10^4}$$

$$= 0.12\ \text{m} = 12\ \text{cm}.$$

10.33. A permanent magnet is required to provide a flux of 0.20 mWb across an air gap of length 0.50 cm and cross-section area 2.0 cm². Calculate the minimum volume of material if the values for the maximum product BH are $B = 0.60$ T and $H = 3.0\times10^4$ At m^{-1}. [44 cm³.]

10.34. The demagnetization characteristic of a permanent magnet is given by the following:

Demagnetizing force (At mm^{-1})	0	8.0	16	24	32	40
Remanent flux density (T)	0.73	0.69	0.61	0.50	0.31	0

Calculate the dimensions of the smallest permanent magnet of this material which will provide a flux density of 0.60 T across an air gap of dimensions 4.0 mm $\times$ 2.0 cm². [7.5 cm $\times$ 2.5 cm².]

Magnetic Moments

10.35. Worked Example

The axis of a bar magnet in air makes an angle of 40° with the direction of a magnetic field in which it lies. The couple exerted on the magnet by the field is 1.0×10^{-6} Nm when the flux density of the field is 1.5×10^{-5} T. Calculate (a) the magnetic moments of the magnet, (b) the pole strength of the magnet if the effective length is 8.0 cm, and (c) the current required in an equivalent air-cored solenoid, having 200 turns of cross-section area 1.0 cm².

Introduction

A suspended magnet behaves very much like a pivoted coil carrying current in a magnetic field (see ex. 1.7 for the moving-coil instrument). In fact the continual spin of electrons in the atoms of the magnet provides the magnet with the equivalent of a current in a coil around it. Thus the long, straight arrow of Fig. 10.14 may represent either a pointer along the axis of a coil, or a suspended bar magnet, turning into line with the flux.

For the coil, the torque or couple due to the forces F has a maximum value T when $\theta = 90°$, given by eqn. (1.7) of ex. 1.7, i.e.

$$T = NIAB \qquad (10.25)$$

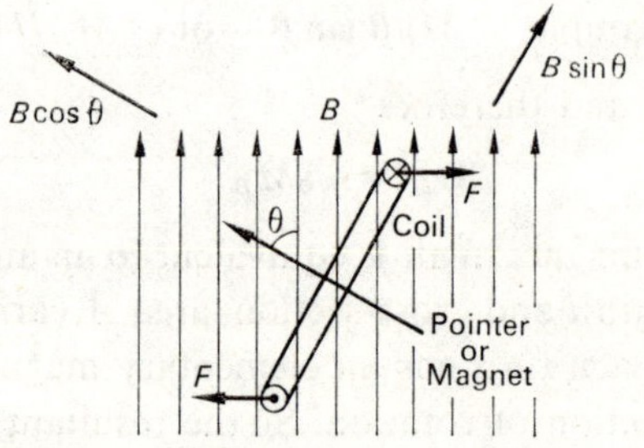

Fig. 10.14. Exs. 10.35 and 10.36.

for N turns of area A, carrying a current I in a field of flux density B, applied perpendicular to the axis of the coil.

Thus for constant current in a particular coil, the maximum value of T is proportional to B, i.e.

$$\frac{T}{B} = NIA = M_B, \tag{10.26}$$

where M_B is called the magnetic B-moment of the coil. The letter B is used to distinguish M_B from the H-moment M_H, which may be defined as the ratio of the maximum value of T to the magnetizing force H of the applied field. Thus, in vacuum,

$$\frac{T}{H} = NIA\mu_0 = M_H. \tag{10.27}$$

Similarly, for the bar magnet, $T \propto B \propto H$, and the magnetic moments M_B and M_H may be defined:

$$M_B = \frac{T}{B} \tag{10.28}$$

and

$$M_H = \frac{T}{H}, \tag{10.29}$$

where B or H is applied perpendicular to the axis of the magnet. When the axis of the coil or magnet makes an angle θ with the direction of the field (as in Fig. 10.14), the effective component of B or H is that perpendicular to the axis. Then

$$\text{the couple} = M_B B \sin\theta \quad \text{or} \quad M_H H \sin\theta, \tag{10.30}$$

where $B = \mu_0 H$ and therefore

$$M_H = \mu_0 M_B. \tag{10.31}$$

A long, thin magnet in air is equivalent to an air-cored solenoid of the same length l and cross-section area A, carrying the appropriate current. Each turn has an elementary magnetic moment IA in the same direction of rotation. So the resultant moment $M_B = NIA$, as for a flat coil.

Now, using eqn. (10.8), the total flux produced by the solenoid in air (or space) is $\varphi_i = \mu_0 \frac{NI}{l} A$.

$$\therefore\ M_B = NIA = \frac{\varphi_i l}{\mu_0} \qquad (10.32)$$

or

$$M_H = NIA\mu_0 = \varphi_i l. \qquad (10.33)$$

This shows that the magnetic moments have two factors: (1) the pole strength φ_i/μ_0 in At m, or φ_i in Wb, where φ_i is the "intrinsic" flux of the magnet or solenoid, as distinct from any applied field B; (2) the effective length l between the poles, being somewhat less than the full length due to some flux passing through the sides of the magnet or solenoid near the poles.

Solution

(a) Statement (10.30) gives $10^{-6} = M_B \times 1.5 \times 10^{-5} \times \sin 40°$.

$$\therefore\ M_B = \frac{10^{-1}}{1.5 \times 0.643} = 0.10 \text{ At m}^2. \qquad \text{(a)}$$

Then $M_H = \mu_0 M_B = 4\pi \times 10^{-7} \times 0.10 = 1.3 \times 10^{-7}$ Wb m. (a)

(b) From eqn. (10.32), the pole strength

$$\frac{\varphi_i}{\mu_0} = \frac{M_B}{l} = \frac{0.10}{8 \times 10^{-2}} = 1.3 \text{ At m} \qquad \text{(b)}$$

or

$$\varphi_i = 1.3 \times 4\pi \times 10^{-7} = 1.6 \times 10^{-6} \text{ Wb}. \qquad \text{(b)}$$

(c) From equation (10.31), the solenoid current

$$I = \frac{M_B}{NA} = \frac{0.10}{200 \times 10^{-4}} = 5 \text{ A}. \qquad \text{(c)}$$

10.36. Worked Example

A small bar magnet of effective length 4.0 cm and moment of intertia 1.0×10^{-6} kg m² produces a magnetic flux of 1.0 μWb. The magnet is suspended as a vibration magnetometer. Calculate the flux density of the magnetic field in air in which the period of vibration is 2.0 s.

Introduction and Solution

In a vibration magnetometer, the magnet is suspended horizontally from a torsionless fibre. It then oscillates when placed in a horizontal field.

Figure 10.14 shows that when the magnet makes an angle θ with the flux of density B, the component of B perpendicular to the axis of the magnet is $B \sin \theta$. Then, from eqn. (10.30), the couple tending to restore θ to zero is $M_B B \sin \theta$. It may be recognized that this indicates that the motion will be simple harmonic and of period $T = 2\pi \sqrt{(I_m/M_B B)}$, where I_m is the moment of inertia of the magnet, and M_B is the magnetic B-moment of the magnet, given by eqn. (10.32).

Thus $$M_B = \frac{\varphi_i l}{\mu_0} = \frac{10^{-6}\times 4\times 10^{-2}}{4\pi\times 10^{-7}} = 3.18\times 10^{-2} \text{ At m}^2.$$

Then by squaring the expression for T,

$$B = \frac{4\pi^2 I_m}{T^2 M_B} = \frac{4\pi^2\times 10^{-6}}{2^2\times 3.18\times 10^{-2}} = 3.1\times 10^{-4} \text{ T}.$$

10.37. A bar magnet, producing a flux of 1.0×10^{-5} Wb, has an effective length of 15 cm. Calculate the couple acting on the magnet when its axis is at an angle of 60° to a magnetic field of flux density 2.5×10^{-5} T in air. [2.6×10^{-5} Nm.]

10.38. A small bar magnet is suspended in air by a thread, with the axis of the magnet horizontal. A vertical magnetic field of 1.0×10^{-5} T is then applied, causing the magnet to dip. The magnet is then brought to the horizontal by loading it with a mass of 200 mg at a point 5.0 mm from the centre. Calculate the magnetic moments of the magnet. [1.2×10^{-6} Wb m; 0.98 At m^2.]

10.39. A solenoid consists of 100 turns of fine wire wound uniformly on a length of 14 cm of a pencil of diameter 7.5 mm. Calculate (a) the magnetic *B*-moment, and (b) the pole strength in webers, when a current of 100 mA flows, assuming the effective length to be the full 14 cm and the relative permeability of the pencil to be unity. [(a) 4.4×10^{-4} At m^2; (b) 4.0×10^{-9} Wb.]

10.40. A vibration magnetometer in the Earth's magnetic field makes 25 vibrations per minute in a horizontal plane. When an additional field of 48 At m^{-1} is applied in the same direction as the horizontal component of the Earth's field, the rate of vibration is doubled. Calculate the Earth's horizontal component. [16 At m^{-1}.]

10.41. A short bar magnet of small cross-section has a magnetic *B*-moment of 1.0 At m^2. Calculate the flux density of the magnetic field at a point on the axis of the magnet at a distance of 40 cm from the centre. Assume that the distance of the point is so much greater than the length of the magnet that the field at the point is the same as if the magnet were a flat coil carrying the appropriate current. [3.1 μT.]

10.42. A small bar magnet having a moment of inertia of 1.0×10^{-6} kg m^2 vibrates a with frequency of 96 vibrations per minute when suspended horizontally as a vibration magnetometer in the magnetic field of the Earth. A magnetic compass is then set in place of the bar magnet, and the latter is set due East of the compass with its axis horizontal and lying East to West. When the distance between the centres of the bar magnet and compass is 35 cm, the compass points North-East. Calculate (a) the horizontal component of the Earth's field, and (b) the magnetic moments of the magnet. Employ the assumption stated in ex. 10.41. All directions are relative to the magnetic North. [(a) 22 μT; (b) 4.7 At m^2 and 5.9 μWb m.]

Susceptibility

10.43. Worked Example

Aluminium has a volume susceptibility of 2.2×10^{-5}, and a density of 2700 kg m^{-3}. Calculate (a) the absolute permeability, and (b) the magnetic moment of a bar of aluminium of mass 1.0 g in a field of magnetizing force 1.0×10^4 At m^{-1}.

Introduction

When a magnetizing force H is applied to a material medium, the resulting flux density B is due not only to H, but also to an "intrinsic" field strength H_i produced by the motion of the electrons in the medium (see also ex. 10.35). The same flux density B would be produced in a vacuum by an applied magnetizing force equal to $(H+H_i)$.

Thus

$$B = \mu_0(H+H_i) = \mu_0 H+\mu_0 H_i, \tag{10.34}$$

where μ_0 = the permeability of free space and $\mu_0 H_i$ = the "intrinsic" flux density B_i, or intensity of magnetization. But

$$B = \mu H = \mu_r \mu_0 H, \tag{10.35}$$

where μ is the absolute permeability of the medium, defined as the ratio of B to H, without reference to H_i, and μ_r is the relative permeability μ/μ_0.

Now, combining eqns. (10.34) and (10.35),

$$\mu_r = \frac{B}{\mu_0 H} = 1+\frac{H_i}{H},$$

where H_i/H is called the "volume susceptibility" $\varkappa$ of the medium, and is dimensionless like μ_r.

Thus

$$\mu_r = 1+\varkappa. \tag{10.36}$$

The numerical values of $\varkappa$ are more convenient than those of μ or μ_r when comparing materials which are only slightly magnetic.

Paramagnetic materials are those with small positive values of $\varkappa$ and H_i. Diamagnetic materials are those with small negative values.

Alternatively, the volume susceptibility may be defined as the ratio B_i/H ($= \mu_0\varkappa$). This gives a different numerical value, and has the units of absolute permeability, henry per metre (H m^{-1}).

The magnetic moment M_B of a rod of the material of length l and cross-section area A is given by eqn. (10.32), i.e.

$$M_B = \frac{\varphi_i}{\mu_0} l, \tag{10.37}$$

where the intrinsic flux

$$\varphi_i = A\mu_0 H_i. \tag{10.38}$$

Then, from eqns. (10.37) and (10.38),

$$H_i = \frac{\varphi_i}{A\mu_0} = \frac{M_B}{Al}, \tag{10.39}$$

i.e. the intrinsic field strength H_i equals the magnetic B-moment M_B per unit volume Al. Similarly, using eqn. (10.33) for the magnetic H-moment M_H, it follows that

$$B_i = \frac{M_H}{Al}. \tag{10.40}$$

The mass susceptibility χ is the volume susceptibility $\varkappa$ divided by the density ϱ of the material. The units of χ are m^3 kg^{-1}, since ϱ is in kg m^{-3} and $\varkappa$ is dimensionless.

Solution

(a) Using eqn. (10.36),

$$\begin{aligned}\mu &= \mu_0\mu_r = \mu_0(1+\varkappa)\\ &= 4\pi\times10^{-7}\ (1+2.2\times10^{-5})\\ &= (4\pi\times10^{-7})+(8.8\pi\times10^{-12})\ \text{H m}^{-1}. \qquad \text{(a)}\end{aligned}$$

Since the orders of magnitude of the terms in brackets are so different, the value of μ is best stated as shown.

(b) From eqn. (10.39) $M_B = H_i A l$, but $\varkappa = \dfrac{H_i}{H}$.

$$\therefore M_B = \varkappa H A l = \varkappa H (m/\varrho)$$

where m is the mass and ϱ is the density.

$$\therefore\ M_B = (2.2\times10^{-5})\times10^4\times\frac{10^{-3}}{2700} = 8.1\times10^{-8}\ \text{At m}^2. \qquad \text{(b)}$$

Then, from eqn. (10.31),

$$M_H = \mu_0 M_B = 4\pi\times10^{-7}\times8.1\times10^{-8} = 10^{-13}\ \text{Wb m}. \qquad \text{(b)}$$

10.44. Calculate the volume susceptibility of (a) a ferromagnetic steel having a relative permeability of 500, (b) platinum which is a paramagnetic material of relative permeability 1.00026, and (c) gold which is a diamagnetic material of relative permeability 0.999964. [(a) 499; (b) 2.6×10^{-4}; (c) -3.6×10^{-5}.]

10.45. The mass susceptibilities of copper and tungsten are -1.0×10^{-9} m^3 kg^{-1} and $+3.6\times10^{-9}$ m^3 kg^{-1} respectively. The densities are 8900 kg m^{-3} and 19,000 kg m^{-3} respectively. Calculate (a) the relative permeability of each material, and (b) the magnetic moments of a volume of 1.0 cm^3 of each material in a field of magnetizing force 1.0×10^3 At m^{-1}. [(a) 0.9999911, 1.000068; (b) -8.9×10^{-9} At m^2 or -1.1×10^{-14} Wb m; $+6.8\times10^{-8}$ At m^2 or $+8.6\times10^{-14}$ Wb m.]

CHAPTER 11

INDUCTORS

Change of Flux Linkage

11.1. Worked Example

A flat coil of diameter 4.0 cm has 100 turns of fine wire. It lies between the poles of an electromagnet with its plane parallel with the pole faces and is pivoted to turn about a diameter. Calculate (a) the average e.m.f. induced in the stationary coil when the flux density of the magnetic field is increased from zero to 2.0×10^{-2} T in a time of 1.0 ms, and (b) the average magnitude of the e.m.f. when the coil is turned over at a rate of 200 rev s^{-1} while the flux density is constant at the value 2.0×10^{-2} T.

Introduction

Whenever there is a change of magnetic flux through a coil or other circuit, an e.m.f. is induced in the circuit. The change of flux may be due to a change of flux density, as in (a), or to movement of the circuit so as to surround more or less flux, as in (b). In both cases Neumann's law applies:

The induced e.m.f. is equal to the rate of change of flux-linkage.

Thus if the circuit consists of a coil of N turns, each linking with the same flux φ, then the induced e.m.f. in volts is given by

$$e = N\frac{d\varphi}{dt}, \qquad (11.1)$$

where $d\varphi/dt$ is the instantaneous rate of change of flux in webers per second. If the flux changes by the amount $\Delta\varphi$ in a time interval Δt, then the average e.m.f.

$$e = N\frac{\Delta\varphi}{\Delta t}. \tag{11.2}$$

Solution

In (a) the maximum flux through the coil = BA, where B = maximum flux density = 2×10^{-2} T, and A = area enclosed by coil

$$= \pi\times(2\times10^{-2})^2 = 4\pi\times10^{-4}\ \text{m}^2.$$

Then the change of flux from zero

$$\Delta\varphi = BA = (2\times10^{-2})\times(4\pi\times10^{-4}) = 8\pi\times10^{-6}\ \text{Wb},$$

and the time interval $\Delta t = 10^{-3}$ s.

Then, from eqn. (11.2), average e.m.f. $= 100\times\dfrac{8\pi\times10^{-6}}{10^{-3}}$

$= 2.5$ V. (a)

(b) When the coil is turned over about an axis, as in Fig. 11.1, the flux through the coil varies from a maximum in the position shown with the plane horizontal to zero when the plane is vertical. This change of flux linkage is the same as in (a), i.e. $\Delta\varphi = 8\pi\times10^{-6}$

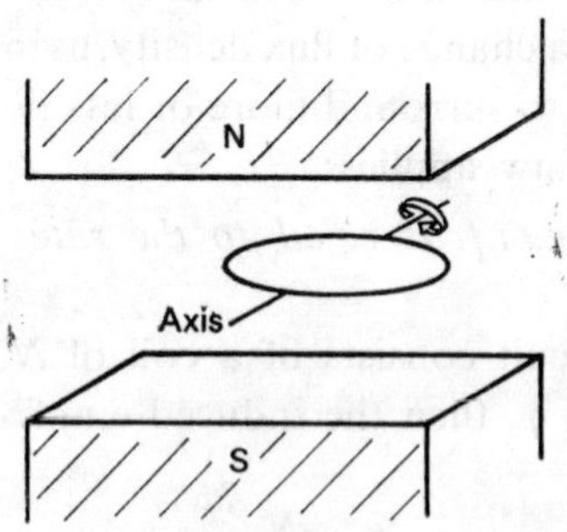

FIG. 11.1. Ex. 11.1.

Wb, but the time interval is now a quarter of the time for one revolution, i.e. $\Delta t = \frac{1}{4}\times\frac{1}{200}$ s. Then

$$\text{average e.m.f.} = N(\Delta\varphi/\Delta t) = 100\times\frac{8\pi\times10^{-6}}{1/800} = 2.0\ \text{V}. \qquad \text{(b)}$$

In each full revolution, four similar changes of flux linkage occur positively or negatively, producing the same average magnitude but not direction of e.m.f.

11.2. The flux through a coil of 50 turns changes from 0.020 mWb to 0.150 mWb in 1.0 ms. Calculate the induced e.m.f. Also calculate the number of turns for a tapping on the coil from which the e.m.f. would be 2.6 V. [6.5 V; 20.]

11.3. A certain change of flux through a coil of 600 turns induces an e.m.f. of 1.5 V. Calculate the number of turns required to obtain an e.m.f. of 3.0 V for the same flux change in a third of the time. [400.]

11.4. A flat coil of 50 turns encloses an area of 10 cm^2 and has a resistance of 20 Ω. A magnetic field of flux density 0.025 T passes through it at right angles to its plane, and collapses to zero in 25 ms. Calculate the average value of induced current on short circuit. [2.5 mA.]

11.5. A coil of 2000 turns links with a flux of 3.0 mWb. Calculate the time in which the flux should be reversed to induce an average e.m.f. of 50 V. [0.24 s.]

11.6. A flat rectangular coil, 3.0×4.0 cm, having 100 turns of fine wire is set rotating at 4500 rev min^{-1} about an axis through its plane and perpendicular to a field of flux density 0.10 T. Calculate (a) the average magnitude of induced e.m.f., and (b) the rate of change of flux density which would produce the same e.m.f. if the coil were at rest with its plane perpendicular to the field. [(a) 3.6 V; (b) 30 T s^{-1}.]

Simple Transformer

11.7. Worked Example

A transformer has a core of cross-section area 3.0 cm² and a primary winding of 720 turns. Calculate (a) the e.m.f. in volts per turn when the flux density in the core varies at the rate of 1.0 T m s^{-1}, (b) the turns ratio for a secondary e.m.f. of 36 V due to the same flux change, and (c) the primary current when the secondary current is 3.0 A.

Introduction

A simple transformer has two windings on the same core. The primary winding carries a current which is varied to change the flux in the core, while the secondary winding receives an induced e.m.f. which provides the output. The variation of flux is the same through each winding (neglecting leakage) and so the same e.m.f. is induced in each turn of each winding.

Thus at any instant

the primary e.m.f. $$e_1 = N_1 \frac{d\varphi}{dt} \qquad (11.3)$$

and the secondary e.m.f. $$e_2 = N_2 \frac{d\varphi}{dt}, \qquad (11.4)$$

where N_1 and N_2 are the numbers of turns of the primary and secondary windings, and $d\varphi/dt$ is the instantaneous rate of change of flux, as in ex. 11.1.

Both e.m.f.s follow Lenz's rule:

The direction of an induced e.m.f. is such as to oppose the change causing it.

Thus the "back e.m.f." e_1 opposes and equals the supply e.m.f. applied to an ideal transformer, while the "output e.m.f." e_2 provides current in such a direction through the secondary winding as to oppose the change of flux in the core.

In an ideal transformer, with negligible reluctance and losses, the primary and secondary instantaneous powers must be equal.

$$\therefore\ e_1 i_1 = e_2 i_2, \tag{11.5}$$

where i_1 and i_2 are the primary and secondary currents at the same instants at which the e.m.f.s are e_1 and e_2.

Then from eqns. (11.3), (11.4), and (11.5) is obtained the "turns ratio"

$$\frac{N_1}{N_2} = \frac{e_1}{e_2} = \frac{i_2}{i_1}. \tag{11.6}$$

Note the inversion of the numbers of the currents in the above ratio.

To distinguish N_1/N_2 from N_2/N_1, the ratio is stated as being "step up" or "step down" when the secondary e.m.f. is greater or less than the primary respectively.

Solution

(a) When B varies at the rate of 1 T per 10^{-3} s through the cross-section area A of 3×10^{-4} m^2,

$$\frac{d\varphi}{dt} = A\frac{dB}{dt} = (3\times10^{-4})\times\left(\frac{1}{10^{-3}}\right) = 0.3 \text{ Wb s}^{-1}.$$

Then, from either eqn. (11.3) or eqn. (11.4), the e.m.f. in volts per turn is

$$\frac{e}{N} = \frac{d\varphi}{dt} = 0.3 \text{ V per turn.} \tag{a}$$

(b) Since e/N is the same for each winding,

$$N_2 = e_2/0.3 = 36/0.3 = 120 \text{ turns.}$$

Then the turns ratio $\dfrac{N_1}{N_2} = \dfrac{720}{120} = 6$,

i.e. the transformer ratio is 6 : 1 step down. (b)

(c) From eqn. (11.6), $i_1 = \dfrac{N_2}{N_1} i_2 = \dfrac{1}{6}\times 3 = 0.5$ A. (c)

11.8. A 4 : 1 step up transformer has a primary winding of 250 turns. Calculate (a) the primary and secondary e.m.f.s when the flux changes at the rate of 0.10 Wb s^{-1}, and (b) the secondary current when the primary current is 300 mA. [(a) 25 V, 100 V; (b) 75 mA.]

11.9. A mains transformer, giving 3 : 1 step up, has a secondary winding of 3000 turns. At one instant, the primary e.m.f. is 240 V and the primary current is 300 mA. Calculate (a) the rate of change of flux in the core, and (b) the secondary e.m.f. and current. [(a) 0.24 Wb s^{-1}; (b) 720 V, 100 mA.]

11.10. The secondary winding of a transformer, with a core of cross-section area 5.0 cm^2, has 600 turns. For a short time the primary winding has a constant applied p.d. of 0.125 volt per turn, and the secondary current is 200 mA. Calculate (a) the rate of change of flux density in the core, (b) the secondary e.m.f., and (c) the m.m.f. of the primary winding. [(a) 250 T s^{-1}; (b) 75 V; (c) 120 At.]

Self-inductance

11.11. Worked Example

A coil of 300 turns is wound on a core having a mean flux path length of 30 cm and cross-section area of 6.0 cm^2. The relative permeability of the material is assumed to have a constant value of 200 for low levels of flux density. Calculate (a) the self-inductance, and (b) the maximum current if the flux density is not to exceed 50 mT.

Introduction

In any coil in which a current produces a magnetic flux, a change of current induces an e.m.f. This e.m.f. e is related to the rate of change of flux $d\varphi/dt$, caused by the change of current, as in eqn.

(11.3):

$$e = N\, d\varphi/dt, \tag{11.7}$$

where N is the number of turns of the coil.

Alternatively, the e.m.f. may be compared with the rate of change of the magnetizing current. Thus

$$e = L\, di/dt, \tag{11.8}$$

where e is in volts, di/dt is the rate of change of current in amperes per second (A s^{-1}), and L is the self-inductance in henrys (H).

Negative signs may be introduced in eqns. (11.7) and (11.8), to emphasize Lenz's rule, but are unnecessary here.

Now eqns. (11.7) and (11.8) may be equated, and integrated if L is a constant:

$$\int N(d\varphi/dt)\, dt = \int L(di/dt)\, dt.$$

$$\therefore\ N\varphi = Li, \tag{11.9}$$

where φ is the flux through the coil when the current is i, and if $\varphi = 0$ when $i = 0$.

But eqn. (10.14) gives

$$Ni = \varphi S \tag{11.10}$$

for a simple coil through which the flux path has reluctance $S = l/\mu A$.

Combining eqns. (11.9) and (11.10) gives

$$L = \frac{N\varphi}{i} = \frac{N^2}{S} \tag{11.11}$$

for a constant reluctance S, as in an air-cored coil.

Solution

(a) Assuming a constant permeability of 200, the reluctance is constant:

$$S = \frac{l}{\mu A} = \frac{0.30}{(200\times 4\pi\times 10^{-7})\times(6\times 10^{-4})} = 2\times 10^6 \text{ At Wb}^{-1}.$$

Then, from eqn. (11.11), the self-inductance

$$L = N^2/S = \frac{300^2}{2\times10^6} = 4.5\times10^{-2}\,\text{H} = 45\text{ mH.} \qquad \text{(a)}$$

(b) The flux density $B = \mu(Ni/l)$.

Therefore if B is the maximum flux density, the maximum current

$$I = \frac{Bl}{\mu N} = \frac{(50\times10^{-3})\times0.30}{(200\times4\pi\times10^{-7})\times300} = 0.2\text{ A.} \qquad \text{(b)}$$

11.12. A flux of 0.40 μWb is produced by a current of 250 mA in an air-cored coil of 1000 turns. Calculate the self-inductance. [1.6 mH.]

11.13. The field winding of a d.c. motor has 1250 turns of self-inductance 20 H and carries a current of 0.50 A. Calculate (a) the total flux, and (b) the back e.m.f. generated when the current is reduced steadily to zero in a time of 10 ms. [(a) 8 mWb; (b) 1000 V.]

11.14. A back e.m.f. of 24 mV is produced in an air-cored coil of 1000 turns when a current of 20 mA is changed by 20% in a time of 1.0 ms. Calculate (a) the rate of change of flux, (b) the self-inductance of the coil, and (c) the self-inductance of each half of the coil between one end and a centre tap. [(a) 2.4×10^{-5} Wb s^{-1}; (b) 6 mH; (c) 1.5 mH.]

11.15. A toroid of 600 turns is wound on a metal ring in which the flux path has a mean length of 20 cm and cross-section area of 3.0 cm^3. When a current of 10 mA is reversed in a time of 10 ms, a mean e.m.f. of 0.54 V is induced. Calculate (a) the self-inductance, and (b) the permeability of the metal, assuming this to be constant. [(a) 0.27 H; (b) 5×10^{-4} H m^{-1}.]

11.16. A ring core has a mean radius of 10 cm, a cross-section area of 10 cm^2, and a winding of 100 turns. Calculate the self-inductance (a) without, and (b) with an air gap of 0.50 mm, assuming a constant relative permeability of 1000 for the core material. [(a) 20 mH; (b) 11 mH.]

Stored Energy

11.17. Worked Example

Calculate the energy stored in the magnetic field of an inductor of inductance 0.20 H when carrying a current of 30 mA.

Introduction

Energy is required to establish the current against the back e.m.f. which occurs as the flux develops. Evidence that energy is stored in the magnetic field is provided by the e.m.f. which appears if the flux is reduced.

Now, if e and i are the average values of applied e.m.f. and current for a very short interval of time δt, during the growth of the flux, the average power is ei, and the energy supplied during this time is $ei\,\delta t$. Since generally both e and i vary, the total energy required is given by

$$W = \int_0 ei\,dt, \qquad (11.12)$$

where e and i are now instantaneous values.

But $e = L(di/dt)$ where di/dt is the rate of change of the current through the coil of inductance L, which is assumed to be constant.

$$\therefore\ W = \int_0^t \left(L\frac{di}{dt}\right) i\,dt = \int_0^I Li\,di = \frac{1}{2}LI^2, \qquad (11.13)$$

where I is the final current.

Solution

When $L = 0.2$ H and $I = 30\times 10^{-3}$ A,

$$W = \tfrac{1}{2}\times 0.2\times(30\times 10^{-3})^2 = 9\times 10^{-5}\ \text{J}.$$

11.18. Calculate the self-inductance of a coil, of resistance 10 Ω, which stores energy of 2.1 J when connected to a supply of e.m.f. 24 V. [0.73 H.]

11.19. A coil of 500 turns is fitted with a core of constant reluctance 4.0×10^5 At Wb^{-1}. Calculate the energy stored by the core when a current of 50 mA is passed through the coil. [7.8×10^{-4} J.]

Mutual Inductance

11.20. Worked Example

Two air-cored coils, of self-inductance 4.0 mH and 9.0 mH respectively, are situated so that their mutual inductance is 3.0 mH. Calculate (a) the e.m.f. induced in one coil when a current of 2.0 A is reversed in the other coil in a time of 50 ms, and (b) the e.m.f. induced in the circuit of the two coils in series, when a current of 2.0 A is reversed in a time of 50 ms.

Introduction

The two coils are so situated that some of the flux produced by the current in one coil links with the other coil. Then, if the current is varied, the flux varies and induces an e.m.f., as in a transformer (see ex. 11.7). Whereas leakage was neglected in a transformer, the leakage flux, which is produced by the primary but fails to link with

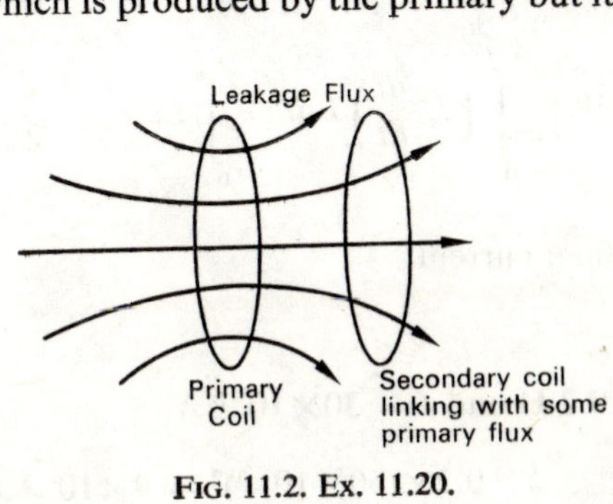

FIG. 11.2. Ex. 11.20.

the secondary, is often considerable in an air-cored mutual inductor (see Fig. 11.2).

Thus the secondary e.m.f. e_2 induced by variation of primary current i_1 at the rate di_1/dt depends not only on the numbers of turns and sizes of the coils but on their relative positions.

Then, for any fixed arrangement of the coils,

$$e_2 = M\frac{di}{dt}, \tag{11.14}$$

where M is the mutual inductance of that arrangement, in henrys (H), when e_2 is in volts, and di_1/dt is in amperes per second. M is the same whichever coil is used as the primary, the other being the secondary. So

$$M = \frac{e_2}{di_1/dt} = \frac{e_1}{di_2/dt}. \tag{11.15}$$

When the coils are connected in series and carry the same varying current, the induced e.m.f. has four components: in coil 1 the e.m.f.s of self and mutual induction are equal to $L(di/dt)$ and $M(di/dt)$ respectively, while in coil 2 the corresponding e.m.f.s are $L_2(di/dt)$ and $M(di/dt)$ again.

Now if the two coils are connected in series aiding, like two parts of one winding, all four component e.m.f.s oppose the change of current. Then the resultant e.m.f. equals

$$(L_1+L_2+2M)\frac{di}{dt}. \tag{11.16}$$

If the connections to one coil or the other are reversed, the direction of any mutually induced e.m.f. is reversed. Then the resultant e.m.f. equals

$$(L_1+L_2-2M)\frac{di}{dt}. \tag{11.17}$$

Hence from eqns. (11.16) and (11.17) the resultant inductance equals

$$L_1+L_2\pm 2M. \tag{11.18}$$

Solution

(a) When a current of 2 A is reversed, the change of current is 4 A. Then, in the interval 50×10^{-3} s, the average rate of change

$$\frac{\Delta i}{\Delta t} = \frac{4}{50\times10^{-3}} = 80\ \text{A s}^{-1}.$$

Now di/dt in eqn. (11.14) refers to the instantaneous rate of change and, if replaced by the average rate, will give the average e.m.f.

Thus the average induced e.m.f. is

$$M\frac{\Delta i}{\Delta t} = (3\times10^{-3})\times80 = 0.24\ \text{V}. \qquad \text{(a)}$$

(b) With the coils in series, either of the resultant self-inductances of eqn. (11.18) may be obtained.

$$\therefore \text{ resultant} \quad L = L_1+L_2\pm2M = 4+9\pm(2\times3)\ \text{mH}$$
$$= 19\ \text{mH} \quad \text{or} \quad 7\ \text{mH}.$$

Then, as in (a), $$\frac{\Delta i}{\Delta t} = 80\ \text{A s}^{-1}.$$

$$\therefore \text{ average induced e.m.f.} = (19\times10^{-3})\times80 \quad \text{or} \quad (7\times10^{-3})\times80$$
$$= 1.5\ \text{V} \quad \text{or} \quad 0.56\ \text{V}. \qquad \text{(b)}$$

11.21. Worked Example

An air-cored mutual inductor of mutual inductance 0.15 mH has a turns ratio of 1 : 5 and coupling coefficient of 100%. Calculate (a) the self-inductance of each winding, and (b) the e.m.f. induced in one winding, when the other is moved to reduce the coupling coefficient to 30%, and is supplied with a pulse of current rising at a constant rate of 4.0 A s^{-1}.

Introduction

The e.m.f. e_2 induced in the secondary is given by two expressions:

(1) from eqn. (11.14), $$e_2 = M\frac{di_1}{dt}, \qquad (11.19)$$

where M is the mutual inductance and di_1/dt is the rate of change of current;

(2) from eqn. (11.1), $$e_2 = N_2\frac{d\varphi_1}{dt}, \qquad (11.20)$$

where $d\varphi_1/dt$ is the rate of change of the primary flux, all of which links with the N_2 turns of the secondary winding when the coupling is 100%.

Now, if M is a constant, eqns. (11.19) and (11.20) may be equated and integrated:

$$\int M\frac{di_1}{dt}\,dt = \int N_2\frac{d\varphi_1}{dt}\,dt.$$

Then $$Mi_1 = N_2\varphi_1, \qquad (11.21)$$

where φ_1 is the flux produced by the primary current i_1, with $\varphi_1 = 0$ when $i_1 = 0$. But eqn. (10.14) gives

$$N_1 i_1 = S\varphi_1, \qquad (11.22)$$

where the primary winding has N_1 turns and S is the constant reluctance of the flux path through both windings.

Combining eqns. (11.21) and (11.22) gives

$$M = \frac{N_1 N_2}{S}. \qquad (11.23)$$

This derivation is similar to that of eqn. (11.11), which gave the self-inductance $L = N^2/S$ for a single coil of N turns.

So, squaring eqn. (11.23) gives $M^2 = \frac{N_1^2}{S}\times\frac{N_2^2}{S} = L_1\times L_2$.

Then, $$M = \sqrt{(L_1 L_2)}, \qquad (11.24)$$

where L_1 and L_2 are the self-inductances of two coils having a mutual inductance M when the coupling is 100%.

If the arrangement is changed so that not all the flux passes through the turns N_2, as in Fig. 11.2, M is reduced by a factor called the coefficient of coupling k. Then

$$M = k\sqrt{(L_1L_2)}. \tag{11.25}$$

Solution

(a) When $\dfrac{N_1}{N_2} = \dfrac{1}{5}$, $\dfrac{L_1}{L_2} = \dfrac{N_1^2/S}{N_2^2/S} = \dfrac{N_1^2}{N_2^2} = \dfrac{1}{25}$.

Then, from eqn. (11.24), $0.15\times10^{-3} = \sqrt{\left(\dfrac{L_2}{25}\times L_2\right)} = \dfrac{L_2}{5}$.

$$\left.\begin{aligned} \therefore\ L_2 &= 0.75\times10^{-3}\ \text{H} = 0.75\ \text{mH} \\ \text{and}\qquad L_1 &= \frac{L_2}{25} = 0.03\ \text{mH}. \end{aligned}\right\} \qquad \text{(a)}$$

(b) When $k = 30\%$, M has this factor as in eqn. (11.25).

$$\therefore\ M = 0.30\times0.15\times10^{-3}\ \text{H} = 0.045\times10^{-3}\ \text{H}.$$

Then, by the definition of M,

$$e_2 = M\frac{di_1}{dt} = 0.045\times10^{-3}\times4$$

$$= 1.8\times10^{-4}\ \text{V} = 0.18\ \text{mV}. \qquad \text{(b)}$$

11.22. When a current of 300 mA is reversed in a coil, an e.m.f. of 2.4 V is induced in another coil in such a position that the mutual inductance is 2 mH. Calculate the time taken to reverse the current, assuming a constant rate of change. [$\frac{1}{2}$ ms.]

11.23. One winding of a mutual inductor is connected to a flux-meter whose calibration constant is 1.0×10^{-4} Wb turns per division. When a current of 1.5 A is reversed in the other winding of the inductor, the fluxmeter throw is 45 divisions. Calculate the mutual inductance. [1.5 mH.]

11.24. A coil of 12 Ω resistance produces a flux of 0.10 mWb when connected to a 24 V supply. A second coil of 200 turns links with 75% of this flux. Calculate (a) the mutual inductance, and (b) the average e.m.f. induced in the second coil when the current is reduced to zero in a time of 5.0 ms. [(a) 7.5 mH; (b) 3 V.]

11.25. A 10 : 1 step up transformer has a primary winding of 200 turns with self-inductance of 0.20 H, carrying a current of 0.25 A which is reversed in a time of 5.0 ms. Calculate (a) the change of flux, (b) the primary induced e.m.f., (c) the secondary induced e.m.f., and (d) the mutual inductance. [(a) $\frac{1}{2}$ mWb; (b) 20 V; (c) 200 V; (d) 2 H.]

11.26. Two coils have self-inductances of 4.0 mH and 9.0 mH respectively. When the coils are connected in series the resultant self-inductance is 22 mH. Calculate (a) the coefficient of coupling, and (b) the resultant self-inductance of the coils when the connections to one coil are reversed. [(a) 75%; (b) 4 mH.]

11.27. A long air-cored solenoid has a diameter of 1.0 cm and 25 turns of wire per centimetre length. A secondary coil of 25 turns is wound over the centre of the solenoid. Calculate the secondary e.m.f. produced by a change of primary current from 100 mA to zero in a time of 1.0 ms. Also calculate the mutual inductance. [0.62 mV; 6.2 μH.]

11.28. Calculate the number of secondary turns for a mutual inductance of 50 μH using a long former of radius 1.0 cm, air-cored, with a primary winding of 20 turns per centimetre. [63.]

11.29. Two coils with a coefficient of coupling of 60% have a resultant self-inductance of 39.4 mH when connected in series. Given the self-inductance of one coil alone to be 16 mH, calculate the possible values of self-inductance of the other coil. [9 mH or 60.8 mH.]

11.30. Two identical coils on the same core have self-inductance of 50.0 mH each and a coupling coefficient of 50%. Calculate the various values of resultant inductance obtainable by connecting the coils to two terminals. [12.5 mH; 37.5 mH; 50 mH; 150 mH.]

Testing Magnetic Material

11.31. Worked Example

A rectangular core is made of laminated material to give a flux path length of 40 cm and cross-section of 2.0×2.0 cm. A primary winding on the core has 500 turns, and a secondary has 5 turns of negligible resistance. A ballistic galvanometer of resistance 120 Ω and sensitivity 5.0 mm μC^{-1} is connected across the secondary, while a d.c. supply is connected to the primary through suitable switches and resistors. When a primary current of 1.6 A is reversed, the galvanometer throw is 15 cm, and when the current of 1.6 A is switched off, the throw is 3.0 cm. Calculate (a) the values of B and H when the current is 1.6 A, and (b) the corresponding value of remanence.

Introduction

The two windings form a mutual inductor in which the change of primary current produces a pulse of secondary e.m.f. At any instant this e.m.f. is given by (11.4): $e_2 = N_2(d\varphi/dt)$, where $d\varphi/dt$ is the rate of change of flux through the secondary of N_2 turns.

Then the current through the galvanometer connected across the secondary,

$$i_2 = \frac{e_2}{R_2} = \frac{N_2}{R_2}\frac{d\varphi}{dt},$$

where R_2 is the total resistance of the secondary circuit.

Now if Q_2 is the total charge circulated by the pulse of e.m.f.,

$$Q_2 = \int_0^\infty i_2\,dt = \int_0^\infty \frac{N_2}{R_2}\frac{d\varphi}{dt}\,dt = \frac{N_2}{R_2}\Delta\varphi, \tag{11.26}$$

where $\Delta\varphi$ is the total change of flux occurring during the pulse, and Q_2 is in coulombs when R_2 is in ohms and $\Delta\varphi$ in webers.

Since Q_2 is measured by the galvanometer, $\Delta\varphi$ can be calculated.

Solution

(a) Since the galvanometer sensitivity is 5 mm μC^{-1}, the throw of 15 cm or 150 mm indicates that the charge Q_2 is $\frac{150}{5} = 30\ \mu C$.

Then, from eqn. (11.26),

$$\Delta\varphi = \frac{R_2 Q_2}{N_2} = \frac{120\times(30\times10^{-6})}{5}$$

$$= 7.2\times10^{-4}\ \text{Wb}.$$

This is the change of flux when the primary current is reversed through zero, causing a change of magnetization from X to Z or

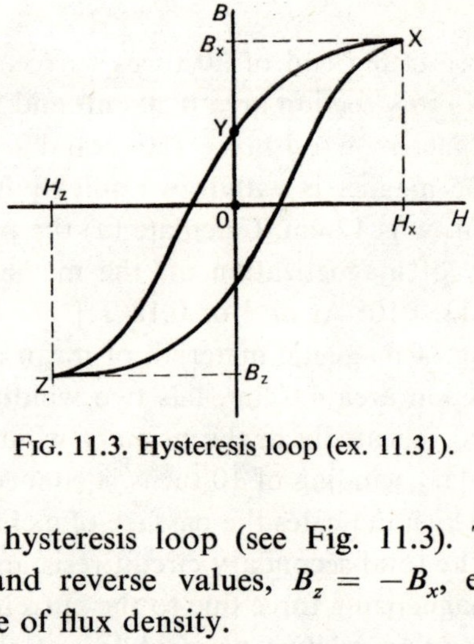

FIG. 11.3. Hysteresis loop (ex. 11.31).

Z to X on the hysteresis loop (see Fig. 11.3). Then, assuming equal forward and reverse values, $B_z = -B_x$, each being half the total change of flux density.

$$\therefore\ B_x = \frac{1}{2}\times\frac{\Delta\varphi}{A},$$

where A is the cross-section area of the core.

$$\therefore\ B_x = \frac{1}{2}\times\frac{7.2\times10^{-4}}{4\times10^{-4}} = 0.9\ \text{T}. \qquad \text{(a)}$$

The corresponding value of H_x is

$$\frac{N_1 I_1}{l} = \frac{500 \times 1.6}{0.4} = 2000 \text{ At m}^{-1}. \qquad \text{(a)}$$

(b) When the primary current is switched from I_1 to zero, the magnetization changes from the tip X to the point of remanence Y in Fig. 11.3. This change from X to Y causes the galvanometer throw of 3.0 cm compared with the previous throw of 15 cm for the change from X to Z.

$$B_x - B_y = \tfrac{3}{15}(B_x - B_z) = \tfrac{3}{15}[0.9 - (-0.9)] = 0.36 \text{ T}.$$

$$\therefore \text{ remanence } B_y = B_x - 0.36 = 0.9 - 0.36 = 0.54 \text{ T}. \qquad \text{(b)}$$

11.32. A close-fitting coil of 20 turns surrounds the centre of a bar magnet of cross-section area 0.50 cm² and is connected to a ballistic galvanometer of resistance 150 Ω and sensitivity 100 mm μC^{-1}. When the magnet is withdrawn quickly from the coil, the galvanometer throw is 12 cm. Calculate (a) the pole strength, and (b) the intensity of magnetization of the magnet. [(a) 7.2 At m or 9 μWb; (b) 1.4×10^5 At m^{-1} or 0.18 T.]

11.33. A ring of magnetic material, of mean diameter 10.0 cm and of cross-section area 4.0 cm², has two windings. A current of 1.0 A is reversed repeatedly in the primary winding of 200 turns, while the secondary winding of 10 turns is connected to a ballistic galvanometer, which indicates the passage of a charge of 30 μC for each reversal. The total secondary circuit resistance is 200 Ω. Calculate (a) the magnetizing force due to the current in the coil, and (b) the corresponding relative permeability of the core material. [(a) 640 At m^{-1}; (b) 940.]

11.34. To determine the hysteresis loop of a magnetic material, ring stampings of mean diameter 10 cm and width 2.0 cm were piled to a thickness of 1.0 cm and wound with 250 primary and 20 secondary turns. The secondary winding was connected to a fluxmeter indicating a change of flux linkage of 1.0×10^{-4} Wbt/div.

Reversals of primary current of 2.5 A gave repeated fluxmeter throws of 112 div. A change of current from 2.5 A to 1.0 A gave a throw of 7 div. A reduction from 2.5 A to zero gave a throw of 26 div. Calculate the *BH* coordinates of three points on the loop: (a) the tip, (b) that for 1 A, (c) the point of remanence. [(a) 1.4 T, 2000 At m^{-1}; (b) 1.2 T, 800 At m^{-1}; (c) 0.75 T, 0.]

CHAPTER 12

CAPACITORS

Capacitance, P.D., and Charge

12.1. Worked Example

A capacitor, charged to a p.d. of 60 V, holds a charge of 300 μC. Calculate (a) the capacitance, (b) the p.d. when the charge is 0.02 C, and (c) the charge when the p.d. is 20 mV.

Introduction

The two plates or strips of metal foil of a simple capacitor may be charged by connecting one to each terminal of a source of e.m.f. The e.m.f. then drives electrons to collect as a negative charge on one plate and to leave an equal positive charge on the other plate. As these charges increase, they oppose further movement of electrons under the action of the e.m.f. Eventually the capacitor holds a charge Q ($+Q$ on one plate and $-Q$ on the other) sufficient to set up a potential difference V exactly balancing the e.m.f. E.

Q is found to be proportional to V or E; i.e.

$$Q = CV, \tag{12.1}$$

where C is the constant called the capacitance of the capacitor. If Q is measured in coulombs and V in volts, eqn. (12.1) gives C in farads (F). 1 μF (microfarad) $= 10^{-6}$ F, 1 pF (picofarad) = 1 μμF (micromicrofarad) $= 10^{-12}$ F.

Note that C is the symbol for capacitance and also C is the abbreviation for coulomb.

Solution

(a) When the charge $Q = 300\ \mu C$ (microcoulomb) $= 300\times 10^{-6}$ C, the p.d. $V = 60$ V (volt).

Then, from eqn. (12.1), capacitance

$$C = \frac{Q}{V} = \frac{300\times 10^{-6}}{60} = 5\times 10^{-6}\ \text{F},$$

$$= 5\ \mu\text{F}. \qquad \text{(a)}$$

(b) When the charge $Q = 0.02$ C with the same capacitance as in (a), eqn. (12.1) gives p.d.

$$V = \frac{Q}{C} = \frac{0.02}{5\times 10^{-6}} = 4000\ \text{V}$$

$$= 4\ \text{kV}. \qquad \text{(b)}$$

(c) When the p.d. $V = 20$ mV $= 20\times 10^{-3}$ V, and the capacitance is still the same, eqn. (12.1) gives charge

$$Q = CV = 5\times 10^{-6}\times 20\times 10^{-3} = 10^{-7}\ \text{C}. \qquad \text{(c)}$$

12.2. Calculate (a) the charge held by a capacitor of 2.0 μF when charged to a p.d. of 300 V; (b) the capacitance of a capacitor which holds a charge of 36 μC at a p.d. of 240 V; (c) the p.d. across a capacitor of 350 pF when holding a charge of 28 μC. [(a) 6×10^{-4} C; (b) 0.15 μF; (c) 80 kV.]

12.3. A certain capacitor holds a charge of 20 μC at a p.d. of 5.0 V. Calculate (a) the p.d. of a second capacitor of four times the capacitance of the first, holding three times the charge, and (b) the charge held by a third capacitor of half the capacitance of the first, at a p.d. of 1 kV. [(a) 3.75 V; (b) 2×10^{-3} C.]

Charging and Discharging

12.4. Worked Example

(a) Calculate how long it would take to raise the p.d. of a 10 μF capacitor from 24 V to 72 V by means of a constant current of 2.0 A. (b) Calculate the final p.d. when the 10 μF capacitor, with initial p.d. of 24 V, is connected in parallel with an uncharged capacitor of 2.0 μF.

Solution

Since the charge held by the capacitor is proportional to the p.d., extra charge ΔQ is required in proportion to any rise of p.d. ΔV.

Thus eqn. (12.1) gives $Q = CV$, where C is the constant capacitance of the capacitor.

$$\therefore \ \Delta Q = C\,\Delta V.$$

(a) When $C = 10\ \mu\text{F} = 10\times10^{-6}$, and $\Delta V = 72-24 = 48$ V,

$$\Delta Q = (10\times10^{-6})\times48 = 4.8\times10^{-4}\ \text{C}.$$

Now the current I in amperes is the rate of flow of charge in coulombs per second; i.e. $I = \Delta Q/\Delta t$, where Δt is the time taken in seconds for the passage of charge ΔQ.

Then

$$\Delta t = \frac{\Delta Q}{I} = \frac{4.8\times10^{-4}}{2} = 2.4\times10^{-4}\ \text{s} = 240\ \mu\text{s}. \qquad \text{(a)}$$

(b) When two capacitors are connected in parallel, the capacitor of higher p.d. will discharge into the other until both have the same p.d. In this process charge is not destroyed but rearranged.

Now all the charge is held initially by the 10 μF capacitor; i.e.

$$Q = C_1V_1 = (10\times10^{-6})\times24 = 2.4\times10^{-4}\ \text{C}.$$

For a common final p.d. V_2, the capacitors hold charges equal to C_1V_2 and C_2V_2.

$$\therefore\ 2.4\times10^{-4} = C_1V_2 + C_2V_2$$
$$= (10\times10^{-6})V_2 + (2\times10^{-6})V_2$$
$$= 12\times10^{-6}V_2.$$
$$\therefore\ V_2 = \frac{2.4\times10^{-4}}{12\times10^{-6}} = 20\text{ V}. \qquad \text{(b)}$$

12.5. Calculate the charge held by a capacitor of 25 μF at a p.d. of 120 V and the average current if it discharges in a time of 10 ms. [3×10^{-3} C; 0.3 A.]

12.6. Calculate the mean current when the p.d. of a capacitor of 4 μF changes from 150 V to 400 V in an interval 2 ms. [0.5 A.]

12.7. A capacitor of capacitance 5.0 μF, charged to a p.d. of 20 V, is connected in parallel with an uncharged capacitor of unknown capacitance. The final p.d. is 25 V. Calculate the unknown capacitance and the final charge of each capacitor. [15 μF; 25 μC; 75 μC.]

12.8. Calculate the common final p.d. when a capacitor of 2.0 μF at a p.d. 24 V is connected in parallel with a second capacitor of 5.0 μF at a p.d. of 32 V. [30 V or 16 V.]

Series Connection

12.9. Worked Example

When a capacitor of unknown capacitance is connected in series with one of 8.0 μF to a supply of e.m.f. 120 V, it develops a p.d. of 96 V. Calculate (a) the unknown capacitance, and (b) the resultant capacitance of the two capacitors.

Introduction

When a source of constant e.m.f. is applied to capacitors connected in series, the same movement of charge occurs all around the circuit, until the total series p.d. of the capacitors becomes equal

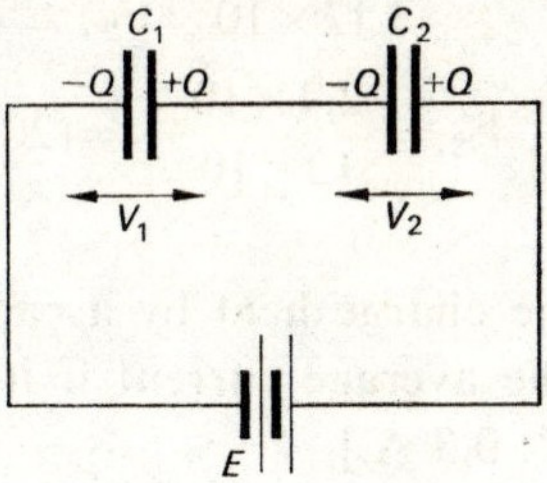

FIG. 12.1. Capacitors in series (ex. 12.1).

to the e.m.f. Thus in Fig. 12.1, electrons, of total charge $-Q$, move clockwise along all three leads until

$$V_1+V_2 = E. \tag{12.2}$$

But from the definition of capacitance,

$$V_1 = \frac{Q}{C_1} \quad \text{and} \quad V_2 = \frac{Q}{C_2}.$$

$$\therefore \frac{Q}{C_1}+\frac{Q}{C_2} = E. \quad \therefore \frac{1}{C_1}+\frac{1}{C_2} = \frac{E}{Q}.$$

Now $E/Q = 1/C$, where C is the capacitance of a single capacitor which would hold the same charge Q when charged by the same e.m.f. E.

Then

$$\frac{1}{C} = \frac{1}{C_1}+\frac{1}{C_2}. \tag{12.3}$$

Similarly,

$$\frac{1}{C} = \frac{1}{C_1}+\frac{1}{C_2}+\frac{1}{C_3} \tag{12.4}$$

for three capacitors in series.

Solution

(a) When C_1 is unknown and $C_2 = 8\ \mu F$, $V_1 = 96$ V for $E = 120$ V.

Then, from eqn. (12.2),

$$V_2 = E - V_1 = 120 - 96 = 24 \text{ V}.$$

Now the charge held by C_2 is $Q = C_2V_2 = (8\times10^{-6})\times24$

$$= 1.92\times10^{-4} \text{ C}.$$

This is equal to the charge held by C_1.

$$\therefore\ C_1 = \frac{Q}{V_1} = \frac{1.92\times10^{-4}}{96} = 2\times10^{-6} \text{ F} = 2\ \mu\text{F}. \qquad \text{(a)}$$

Note that C_1 and C_2 in series act as a potential divider, with the higher p.d. across the smaller capacitor, since $Q = C_1V_1 = C_2V_2$.

(b) The resultant capacitance C can be obtained from eqn. (12.3):

$$\frac{1}{C} = \frac{1}{C_1} + \frac{1}{C_2} = \frac{1}{2} + \frac{1}{8},$$

for C in microfarads.

$$\therefore\ C = \frac{2\times8}{2+8} = 1.6\ \mu\text{F}. \qquad \text{(b)}$$

Alternatively, C may be calculated by using the value of Q obtained in part (a):

$$C = \frac{Q}{E} = \frac{1.92\times10^{-4}}{120} = 1.6\times10^{-6} \text{ F}$$

$$= 1.6\ \mu\text{F}. \qquad \text{(b)}$$

12.10. Two capacitors of capacitance 2.0 μF and 3.0 μF respectively are connected (a) in series, and (b) in parallel to a battery of e.m.f. 300 V. In each case calculate the charge and p.d. of each capacitor. [(a) 360 μC, 180 V, 120 V; (b) 600 μC, 900 μC, 300 V.]

12.11. Three capacitors of capacitance 2.0 μF, 4.0 μF, and 6.0 μF respectively are connected across a supply of e.m.f. 110 V. Calculate the charge and p.d. of each capacitor (a) when connected in parallel, and (b) when connected in series. [(a) 220 μC, 440 μC, 660 μC, 110 V; (b) 120 μC, 60 V, 30 V, 20 V.]

12.12. What capacitor in series with one of 2.0 μF will limit the full charge of the latter to 25 μC on a supply of e.m.f. 20 V? [3.3 μF.]

Series–Parallel Connections

12.13. Worked Example

Two capacitors are connected in series across a source of constant e.m.f. An electrostatic voltmeter having a capacitance of 20 pF is connected first across one capacitor, then across the other, and, finally, across the source. The three voltmeter indications are respectively 600 V, 1000 V, and 2000 V. Calculate the capacitance of the capacitors.

Solution

When two capacitors of capacitance C_x and C_y respectively are connected in parallel, they share a common p.d. V and hold charges VC_x and VC_y, making a total charge $Q = V(C_x + C_y)$. Then the resultant capacitance

$$\frac{Q}{V} = C_x + C_y. \tag{12.5}$$

Thus with the voltmeter in the first position, as in Fig. 12.2, there is a p.d. $V_1 = 600$ V across the parallel resultant $(C_1 + C_v)$ where C_v is the voltmeter capacitance $= 20$ pF $= 20 \times 10^{-12}$ F.

Now the final voltmeter indication shows the e.m.f. E to be 2000 V.

Then $V_2 = E - V_1 = 2000 - 600 = 1400$ V.

Now the total Q of C_1 and C_v in parallel is the same as that of C_2 in series with them.

$$\therefore\ Q = V_2C_2 = V_1(C_1 + C_v).$$

$$\therefore\ 1400C_2 = 600[C_1 + (2\times10^{-11})]. \qquad (12.6)$$

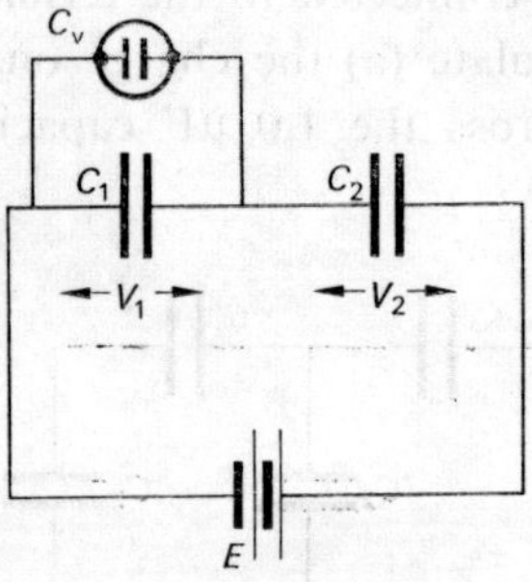

FIG. 12.2. Voltmeter in first position (ex. 12.13).

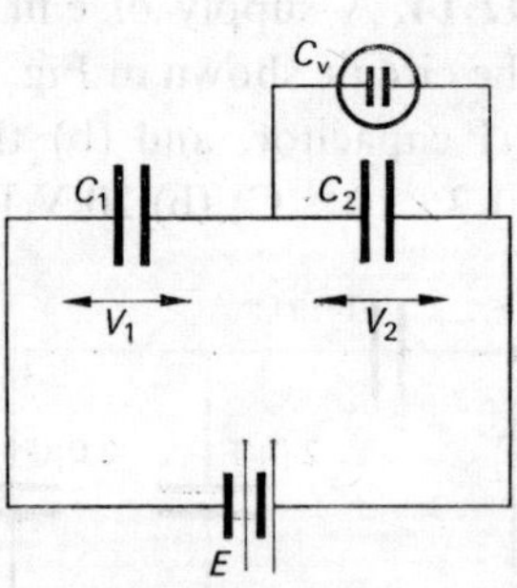

FIG. 12.3. Voltmeter in second position (ex. 12.13).

The second voltmeter indication shows that $V_2 = 1000$ V across the parallel combination of $C_2 + C_v$, as in Fig. 12.3. Then

$$V_1 = E - V_2 = 2000 - 1000 = 1000 \text{ V},$$

and

$$Q = V_1C_1 = V_2(C_2 + C_v).$$

$$\therefore 1000C_1 = 1000\,[C_2 + (2\times10^{-11})]. \qquad (12.7)$$

Equations (12.6) and (12.7) are now to be solved simultaneously: Multiplying eqn. (12.7) by 1.4 gives

$$1400C_1 = 1400C_2 + 2.8\times10^{-8},$$

and rearranging eqn. (12.6) gives

$$600C_1 = 1400C_2 - 1.2\times10^{-8}.$$

Subtracting, $$800C_1 = 4.0\times10^{-8}.$$

$$\therefore\ C_1 = \frac{4.0\times10^{-8}}{800} = 0.50\times10^{-10}\text{ F}$$

$$= 50\text{ pF}.$$

Then, rearranging eqn. (12.7), $1000C_2 = 1000C_1 - 2\times10^{-8}$.

$$\therefore\ C_2 = C_1 - 20\times10^{-12}$$

$$= (50\times10^{-12}) - (20\times10^{-12}) = 30\times10^{-12}$$

$$= 30\ \text{pF}.$$

12.14. A supply of e.m.f. 24 V is connected to the terminals of the circuit shown in Fig. 12.4. Calculate (a) the charge on the 3.0 μF capacitor, and (b) the p.d. across the 1.0 μF capacitor. [(a) 1.2×10^{-5} C; (b) 20 V.]

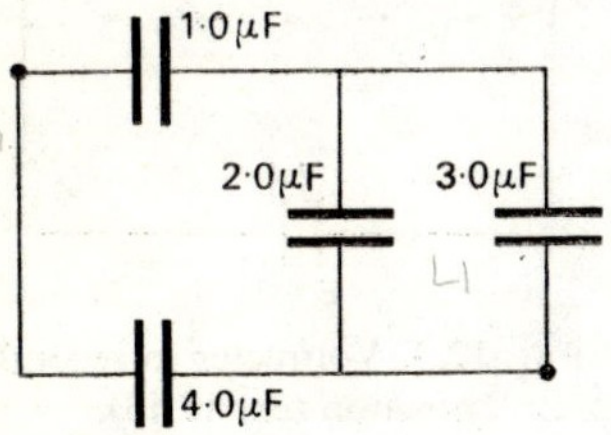

FIG. 12.4. Ex. 12.14.

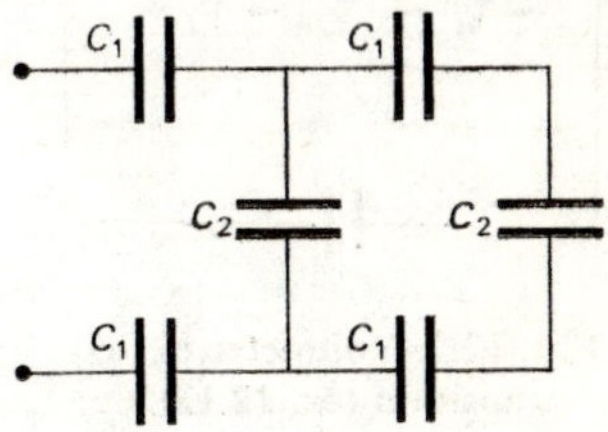

FIG. 12.5. Ex. 12.15.

12.15. Calculate the resultant capacitance of the arrangement shown in Fig. 12.5 where four capacitors have capacitance $C_1 = 0.020$ μF, and two have $C_2 = 0.010$ μF. (0.0060 μF).

12.16. A supply of e.m.f. 110 V is connected to the terminals of the group of capacitors shown in Fig. 12.6, where $C_1 = C_3 = 0.020$ μF, and $C_2 = C_4 = 0.050$ μF. Calculate (a) the charge held by capacitor C_1, and (b) the p.d. across capacitor C_4. [(a) 0.37 μC; (b) 26 V.]

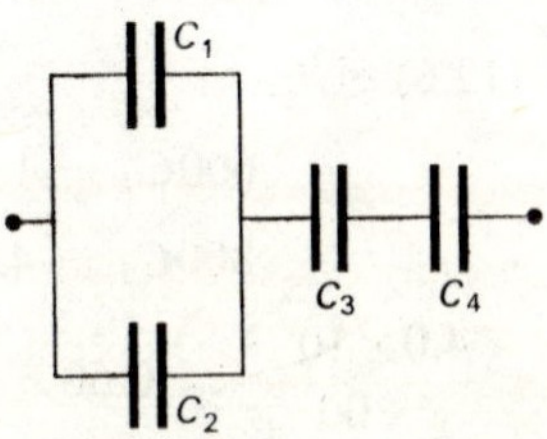

FIG. 12.6. Ex. 12.16.

12.17. Calculate all values of resultant capacitance which may be obtained by connecting the following capacitors, using all three, and using only two at a time: 1 μF, 2 μF, and 4 μF. [$\frac{4}{7}$, $\frac{2}{3}$, $\frac{4}{5}$, $\frac{6}{7}$, $1\frac{1}{3}$, $1\frac{3}{7}$, $1\frac{5}{7}$, $2\frac{1}{3}$, $2\frac{4}{5}$, 3, $4\frac{2}{3}$, 5, 6, 7, all in μF.]

Stored Energy

12.18. WORKED EXAMPLE

Calculate the energy lost when a capacitor of 2.00 μF, at a p.d. of 300 V, is connected across another of 4.00 μF at a p.d. of 200 V. The terminals connected are (a) of like polarity, and (b) of unlike polarity.

Introduction

Energy is stored in a capacitor when it is charged by an e.m.f. The source of e.m.f. does work in driving the charge in the charging circuit against the opposing attractions and repulsions of the charges which build up on the capacitor plates. If the capacitor is discharged through a resistor, energy is released as heat in this resistor.

Now if the charging current is maintained constant at a value I for a time t, the p.d. will rise steadily from 0 to V, with average value $V/2$.

Then the average power = (constant current)×(average p.d.)

$$= I \times \frac{V}{2}$$

and energy supplied and stored $= W$

$$= \text{(average power)} \times \text{time}$$

$$= \tfrac{1}{2} IV \times t.$$

But It = final charge Q of the capacitor.

$$\therefore\ W = \tfrac{1}{2} QV.$$

This may be expressed in terms of the capacitance C, since $C = Q/V$.

Thus,

$$W = \frac{1}{2}QV = \frac{1}{2}CV^2 = \frac{1}{2}\frac{Q^2}{C}, \tag{12.8}$$

where W = stored energy in joules, Q = stored charge in coulombs, V = potential difference in volts, and C = capacitance in farads.

It may be shown that the relationships (12.8) apply even when the charging current is not constant.

Solution

For the first capacitor of the question,

$$W_1 = \tfrac{1}{2}C_1V_1^2 = \tfrac{1}{2}(2\times10^{-6})(300)^2 = 0.090 \text{ J}$$

and for the second,

$$W_2 = \tfrac{1}{2}C_2V_2^2 = \tfrac{1}{2}(4\times10^{-6})(200)^2 = 0.080 \text{ J.}$$

$$\therefore \text{ total energy stored initially} = W_1+W_2 = 0.170 \text{ J.} \tag{12.9}$$

Also the initial charges of the capacitors are:

$$Q_1 = V_1C_1 = 300\times2\times10^{-6} = 6\times10^{-4} \text{ C} \tag{12.10}$$

and

$$Q_2 = V_2C_2 = 200\times4\times10^{-6} = 8\times10^{-4} \text{ C.} \tag{12.11}$$

When the capacitors are connected in parallel (either way) they have a resultant capacitance

$$C = C_1+C_2 = (2\times10^{-6})+(4\times10^{-6})$$
$$= 6\times10^{-6} \text{ F.} \tag{12.12}$$

(a) With terminals of like polarity connected, the total charge

$$Q = Q_1+Q_2 = 14\times10^{-4} \text{ C.} \tag{12.13}$$

This is redistributed without loss of charge to give a common p.d. as if all the charge Q were held by the equivalent capacitor C.

Then the final energy stored is given by eqn. (12.8) with eqns. (12.12) and (12.13):

$$W = \frac{1}{2}\frac{Q^2}{C} = \frac{1}{2}\frac{(14\times10^{-4})^2}{6\times10^{-6}} = 0.163 \text{ J.}$$

This is less than eqn. (12.9), showing a loss of energy of

$$0.170-0.163 = 0.007 \text{ J.} \qquad \text{(a)}$$

(b) With terminals of opposite polarity connected, the charges Q_1 and Q_2 partially neutralize each other. The remaining charge is then $+Q_1-Q_2$ on one side of the resultant capacitor, and $-Q_1+Q_2$ on the other. So the final stored energy is given by eqn. (12.8): $W = \frac{1}{2}Q^2/C$, where from eqns. (12.10) and (12.11),

$$Q = (8\times10^{-4})-(6\times10^{-4}) = 2\times10^{-4} \text{ C}$$

and from eqn. (12.12), $\quad C = 6\times10^{-6}$ F.

$$\therefore\ W = \frac{1}{2}\frac{(2\times10^{-4})^2}{6\times10^{-6}} = 0.003 \text{ J.}$$

$$\therefore \text{ loss of energy} = 0.170-0.003 = 0.167 \text{ J.} \qquad \text{(b)}$$

12.19. A capacitor of 8 μF is charged to a p.d. of 10 V, and then connected across a 2 V accumulator with like polarities together. Calculate the energy lost by the capacitor. [4×10^{-4} J.]

12.20. Calculate how many capacitors, each of capacitance 16 μF, charged to a p.d. of 1000 V, would be needed to store one new penny worth of energy. Take the cost at the rate of $1\frac{1}{2}$ p. per kWh. [3×10^5.]

12.21. Three capacitors, of capacitance 2.0 μF, 3.0 μF, and 6.0 μF respectively, are connected (a) in parallel, and (b) in series, with a source of e.m.f. of 120 V. Calculate the total energy stored in each case. [(a) 0.079 J; (b) 0.0072 J.]

12.22. A capacitor is connected directly across a switch which breaks a current in an inductive circuit of self-inductance 2 H. Calculate the minimum capacitance required when breaking a current of 0.3 A if the p.d. of the capacitor is not to rise above 300 V. [2 μF.]

CHAPTER 13

ELECTRIC FIELDS

Inverse Square Law

13.1. Worked Example

A small sphere carrying a charge of $+4.0\times10^{-9}$ C is situated in air at a point A, 10 cm from a point B.

(a) Calculate (1) the electric flux density at B, (2) the electric field strength at B, and (3) the force exerted on a small sphere carrying a charge of $+2.0\times10^{-9}$ C at B. (b) Calculate the quantities (1), (2), and (3) when an additional third sphere, carrying a charge of -8.0×10^{-9} C, is set at a point C, 10 cm from both spheres A and B.

Introduction

Coulomb's law states that the force F between two point charges Q_1 and Q_2 in a vacuum is proportional to their product and inversely proportional to the square of their distance apart r.

Thus

$$F \propto \frac{Q_1 Q_2}{r^2}. \tag{13.1}$$

Each charge is said to produce an electric field, the strength of which, at any point, is measured by the force exerted per unit charge at that point. Thus at the position of Q_2 the electric field strength E_1 due to Q_1 is F/Q_2 in newtons per coulomb.

Then, using relationship (13.1),

$$E_1 = \frac{F}{Q_2} \propto \frac{Q_1}{r^2}. \tag{13.2}$$

This use of E for electric field strength must not be confused with E for e.m.f. in volts (see also ex. 13.6).

As a visual and mathematical aid to understanding how one charge affects another at a distance, a positive charge at a point is imagined to issue electric flux symmetrically in all directions in a vacuum or homogeneous medium. Similarly, flux converges on negative charges. In M.K.S. units the coulomb is used as the unit for both charge and electric flux.

As the flux from a positive charge diverges outward through spheres of increasing surface area, so the flux density decreases. Thus the flux density D in coulombs per square metre, at a distance r from a point charge Q is given by

$$D = \frac{\text{flux}}{\text{area of sphere}} = \frac{Q}{4\pi r^2}. \tag{13.3}$$

This follows the inverse square law, but the ratio D/E, called the permittivity ε, depends upon the particular medium in which the field of flux is produced.

Then

$$E = \frac{D}{\varepsilon} = \frac{Q}{4\pi\varepsilon r^2}, \tag{13.4}$$

where E is the electric field strength in newton per coulomb, (N C^{-1}), D is the electric flux density in coulomb per square metre (C m^{-2}), Q is the charge in coulomb (C) producing the field, r is the radial distance in metres (m), ε is the permittivity of the medium in farads per metre (F m^{-1}).

For vacuum and air, $\varepsilon_0 = 8.85 \times 10^{-12}$ F m^{-1}. For dielectric materials ε is greater than ε_0 by a dimensionless factor ε_r called the relative permittivity, so that

$$\varepsilon = \varepsilon_r \times \varepsilon_0. \tag{13.5}$$

For explanation of the units of ε and ε_0, see ex. 13.11.

Solution

It is assumed that the spheres are so small that they may be considered to be points.

(a1) When $Q_A = +4\times10^{-9}$ C and $r = 0.1$ m, the flux density at B, due to Q_A, is directed away from A, and is given by eqn. (13.3):

$$D_A = \frac{Q_A}{4\pi r^2} = \frac{4\times10^{-9}}{4\times\pi\times10^{-2}} = 3.2\times10^{-8}\ \text{C m}^{-2}. \tag{a1}$$

(a2) The field strength at B is given by eqn. (13.4):

$$E_A = \frac{D_A}{\varepsilon_0} = \frac{3.2\times10^{-8}}{8.85\times10^{-12}} = 3.6\times10^{3}\ \text{N C}^{-1}. \tag{a2}$$

(a3) The force on charge Q_B of 2×10^{-9} C is given by eqn. (13.2):

$$F_A = E_A Q_B = (3.6\times10^{3})\times(2\times10^{-9})$$
$$= 7.2\times10^{-6}\ \text{N}. \tag{a3}$$

(b1) The flux density D_C, due to the negative charge Q_C, is directed toward C and is given by eqn. (13.3):

$$D_C = \frac{Q_C}{4\pi r^2} = \frac{(-)8\times10^{-9}}{4\pi\times10^{-2}} = (-)6.4\times10^{-8}\ \text{C m}^{-2},$$

where the negative sign shows the force to be one of attraction. The resultant flux density is found by applying the cosine rule to the parallelogram of vectors with the 60° angle shown in Fig. 13.1. Thus

$$D^2 = D_A^2 + D_C^2 - 2D_A D_C \cos 60°$$
$$= [3.2^2 + 6.4^2 - (2\times3.2\times6.4\times0.5)]10^{-16}$$
$$= 30.7\times10^{-16}.$$
$$\therefore\ D = 5.5\times10^{-8}\ \text{C m}^{-2}. \tag{b1}$$

(b2) The resultant field strength

$$E = D/\varepsilon_0 = \frac{5.5\times10^{-8}}{8.85\times10^{-12}} = 6.2\times10^3 \text{ N C}^{-1}. \qquad \text{(b2)}$$

(b3) The resultant force

$$F = EQ_B = (6.2\times10^3)\times(2\times10^{-9}) = 1.2\times10^{-5} \text{ N}. \qquad \text{(b3)}$$

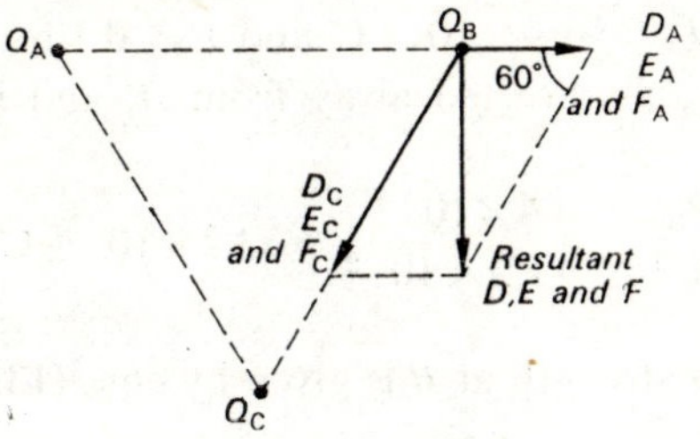

FIG. 13.1. EX. 13.1.

Since $D_C = 2D_A$, the parallelogram of vectors shows D to be perpendicular to D_A. Also E and F follow the same direction as D.

13.2. Calculate the force of attraction between an α-particle, having a charge of $+3.2\times10^{-19}$ C, and an electron having a charge of -1.6×10^{-19} C, when they are separated by a distance of 1.0 μm in vacuum. [4.6×10^{-16} N.]

13.3. Two small spheres, each of mass 20 mg, are suspended in air by threads each of length 50 cm and of negligible weight, attached to the same point. When the spheres are charged equally they repel each other to a distance of 10 cm. Calculate the charge of each sphere. [4.7×10^{-9} C.]

13.4. Four small spheres, A, B, C, and D, carrying charges of 1.0×10^{-9} C, -2.0×10^{-9} C, $+3.0\times10^{-9}$ C, and -4.0×10^{-9} C respectively, are set in cyclic order at the corners of a square of side 20 cm. Calculate the electric field strength in magnitude and direction at the centre of the square when the medium has a relative permittivity of 2.5. [510 N C^{-1} toward mid-point of AD.]

13.5. Two point charges of 1.0×10^{-9} C and 1.0×10^{-8} C respectively are 30 cm apart (a) in vacuum, and (b) in a medium of relative permittivity 2.5. Calculate the force in milligram force exerted by each charge on the other, and the force exerted by both charges on a point charge of 1.0×10^{-10} C situated midway between the two. [(a) 0.10 mgf, 0.037 mgf; (b) 0.041 mgf, 0.015 mgf.]

Potential

13.6. Worked Example

Two small spheres having charges of $+1.0\times10^{-8}$ C and -4.0×10^{-9} C respectively are 20 cm apart in a medium of relative permittivity 4.7. Calculate (a) the work required to increase the spacing to 100 cm, and (b) the absolute potential at the point halfway between them after separation.

Introduction

The potential difference, in volts, between two points in an electric field is defined as the work per unit charge, in joules per coulomb, required to move a positive charge from one point to the other. Now if work δW is required to move a charge Q between two points δx apart, against a force F due to the electric field, $\delta W = -F\,\delta x$, where the negative sign shows work to be done against F, and F is assumed to be constant over the small distance δx.

Then, by definition, the p.d.

$$\delta V = \frac{\delta W}{Q} \tag{13.6}$$

$$= \frac{-F\,\delta x}{Q} = -E\,\delta x, \tag{13.7}$$

where E is the electric field strength defined as in ex. 13.1. Rearranging eqn. (13.7) and reducing δx toward a point gives

$$E = -\frac{dV}{dx} \tag{13.8}$$

which shows that the electric field strength E is numerically equal to the potential gradient dV/dx, and the negative sign shows that the field strength is positive down the potential gradient, as in a gravitational field. Equation (13.8) shows that E may be measured in volts per metre, which are the units of dV/dx, as an alternative to the original units newton per coulomb. This use of E must be distinguished from the use to denote e.m.f. as defined in ex. 1.10.

The absolute potential V at any point is the potential difference between that point and infinity, which is taken as absolute zero of potential. Thus, by integrating eqn. (13.8), $V = -\int_{\infty}^{X} E\,dx$.

Now in the field of a point charge Q, where x is measured along the radius, $x = r$, and $E = Q/4\pi\varepsilon r^2$ from eqn. (13.4).

Then in this situation

$$V = -\int_{\infty}^{r} \frac{Q}{4\pi\varepsilon r^2}\,\mathrm{d}r = \frac{Q}{4\pi\varepsilon r}. \tag{13.9}$$

Solution

(a) In the problem, one charge may be considered to be stationary and the other to be moved. Thus assuming a fixed charge $Q_A = +10^{-8}$ C, the two positions of the other charge Q_B are $r_1 = 0.2$ m and $r_2 = 1.0$ m, and the work to be done is given by eqn. (13.6): $\delta W = Q_B(V_2 - V_1)$, where V_1 is the potential at position r_1, and V_2 at r_2.

Now, from eqn. (13.7),

$$V_1 = \frac{Q_A}{4\pi\varepsilon r_1} \quad \text{and} \quad \varepsilon = \varepsilon_r \varepsilon_0$$

$$\therefore\ V_1 = \frac{10^{-8}}{4\pi(4.7\times 8.85\times 10^{-12})0.2} = 96\ \text{V}.$$

Similarly, $$V_2 = \frac{10^{-8}}{4\pi(4.7\times 8.85\times 10^{-12})1.0} = 19\text{ V}.$$

Then the work required,

$$\delta W = Q_B(V_2 - V_1) = -4\times 10^{-9}(19-96) = 3.1\times 10^{-7}\text{ J.} \quad \text{(a)}$$

(b) The potential at any point in the resultant field of two charges is the algebraic sum of the potentials due to the separate charges. Thus at a distance of 0.5 m from Q_A

$$V_A = \frac{Q}{4\pi\varepsilon r} = \frac{10^{-8}}{4\pi(4.7\times 8.85\times 10^{-12})0.5} = 38\text{ V}.$$

Similarly, at the same point, but due to Q,

$$V_B = \frac{Q}{4\pi\varepsilon r} = \frac{-4\times 10^{-9}}{4\pi(4.7\times 8.85\times 10^{-12})0.5} = -15\text{ V}.$$

Then the resultant potential at the mid point $= 38-15$

$= 23$ V. (b)

13.7. With reference to ex. 13.2, calculate the energy required to move the electron from the distance of 1.0 μm (a) to a point distant 10 μm from the α-particle, and (b) to infinity. [(a) 4.1×10^{-22} J; (b) 4.6×10^{-22} J.]

13.8. With reference to ex. 13.4, calculate the resultant potential at the centre of the square. [−51 V.]

13.9. Calculate (a) the separate amounts of work to be done against gravity and against the electric field of the Earth in raising an α-particle through a height of 5.0 cm near the Earth's surface, and (b) the mass of a particle which has a charge of one electron and which would be supported by the electric field of the Earth acting against gravity near the Earth's surface. Assume (1) the strength of the electric field is 300 V m^{-1} vertically downward, (2) the strength of the gravitational field is given by the acceleration of 9.8 m s^{-2}, (3) the mass of an α-particle is 6.7×10^{-27} kg and its charge is $+3.2\times10^{-19}$ C, (4) the charge of one electron is 1.6×10^{-19} C. [3.3×10^{-27} J; 4.8×10^{-18} J; 4.9×10^{-18} kg.]

13.10. Two small spheres with centres A and B, 1.0 m apart in vacuum, have charges of $+1.0\times10^{-9}$ C and $+2.0\times10^{-9}$ C respectively. Calculate (a) the field strength at the point C midway between A and B, (b) the potential at C, (c) the energy required to move a small sphere having a charge of -2.0×10^{-12} C from C to a point 2.0 m from both A and B. [(a) 36 V m^{-2}; (b) 54 V; (c) 8.1×10^{-11} J.]

Parallel Plates

13.11. Worked Example

A simple capacitor of capacitance 400 pF consists of two parallel plates separated by 2.0 mm of dielectric material. Each face of each plate has an area of 0.040 m^2, and the adjacent faces carry opposite charges of 1.2×10^{-8} C. Calculate (a) the strength of the electric field in the dielectric, and (b) the relative permittivity of the dielectric.

Introduction

It can be shown that the resultant flux near a uniformly charged flat plate of metal is perpendicular to the surface. It does not incline one way or the other except near the edges of the plate or at a distance from it, as in Fig. 13.2. Then if two such plates carrying equal and opposite charges are set parallel at a small distance, each is in the perpendicular field of the other. These two fields of flux, shown in Fig. 13.3, have zero resultant on either side but add in the same direction between the plates, and may be assumed uniform.

Now if the plates have charges $+Q$ and $-Q$, each has half its flux, $\frac{1}{2}Q$, on each side of it, and the resultant flux between the plates is $\frac{1}{2}Q+\frac{1}{2}Q=Q$. Then, if A is the area of each face, the flux density between the plates is

$$D = \frac{Q}{A}. \qquad (13.10)$$

Since D is uniform, so is the field strength or potential gradient.

$$\therefore\ E = \frac{V}{d}, \qquad (13.11)$$

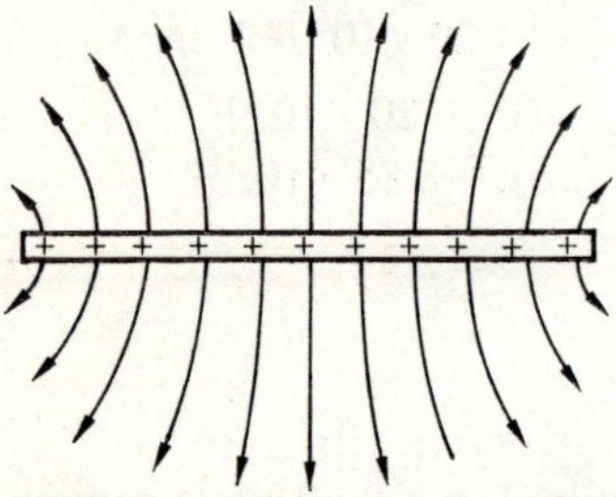

FIG. 13.2. Electric field near a flat charged plate (ex. 13.11).

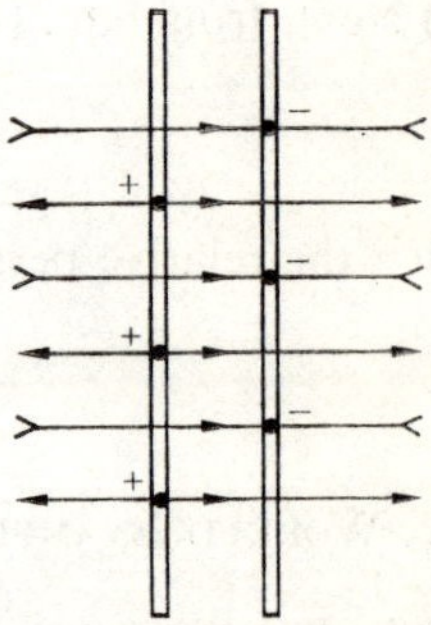

FIG. 13.3. Electric field near two flat parallel plates (ex. 13.11).

where V is the p.d. between the plates and d is their separation. But $D/E = \varepsilon$.

$$\therefore\ \varepsilon = \frac{Q}{A} \div \frac{V}{d} = \frac{Qd}{AV}.$$

Then the capacitance $C = Q/V$ (see eqn. 12.1)

$$= \frac{\varepsilon A}{d}. \qquad (13.12)$$

Rearrangement of this equation shows the units of ε (or ε_0): $\varepsilon = (Cd/A)$, having units of (farads$\times$metres)/metres^2 $\equiv$ farads/metres, i.e. F/m or F m^{-1}.

Solution

(a) From above, $V = \dfrac{Q}{C} = \dfrac{1.2\times10^{-8}}{400\times10^{-12}} = 30$ V.

Then from eqn. (13.11), $E = \dfrac{V}{d} = \dfrac{30}{2\times10^{-3}} = 1.5\times10^{4}$ V m^{-1}

(a)

(b) Next, from eqn. (13.12), $\varepsilon = \dfrac{Cd}{a} = \dfrac{(400\times10^{-12})(2\times10^{-3})}{0.040}$

$= 20\times10^{-12}$ F m^{-1}.

Then the relative permittivity $\varepsilon_r = \dfrac{\varepsilon}{\varepsilon_0} = \dfrac{20\times10^{-12}}{8.85\times10^{-12}}$

$= 2.3.$ (b)

13.12. Worked Example

Calculate the number of plates required in a multiplate capacitor having rectangular plates 2.0×3.0 cm, to obtain a capacitance of 0.0025 μF, when the dielectric material has a thickness of 0.25 mm and a relative permittivity of 7.0.

Introduction

If numerous parallel plates are interleaved with dielectric and connected alternately to two terminals, as in Fig. 13.4, there are, in effect, numerous simple capacitors in parallel. Each simple capacitor consists of one sheet of dielectric between the lower face L of one plate and the upper face U of another plate.

Thus for 4 plates, the 3 equivalent simple capacitors are shown in an equivalent connection in Fig. 13.5. Since the uppermost face U_1 and the lowest face L_4 do not face each other, they provide no capacitance.

Similarly, for N plates there are in effect $N-1$ capacitors, each acting across one of the $N-1$ sheets of dielectric.

Now, from eqn. (13.12) for one capacitor, the capacitance in farads is $C = (\varepsilon A/d)$, where ε is the permittivity of the dielectric in M.K.S. units, A is the area of each face in square metres, and d is the spacing in metres.

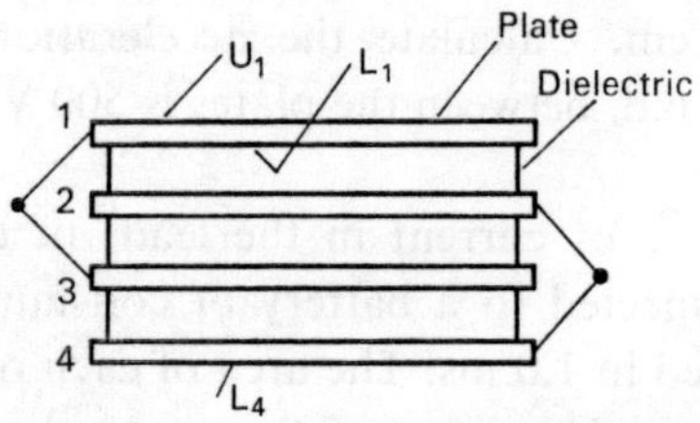

FIG. 13.4. Multiplate capacitor (ex. 13.12).

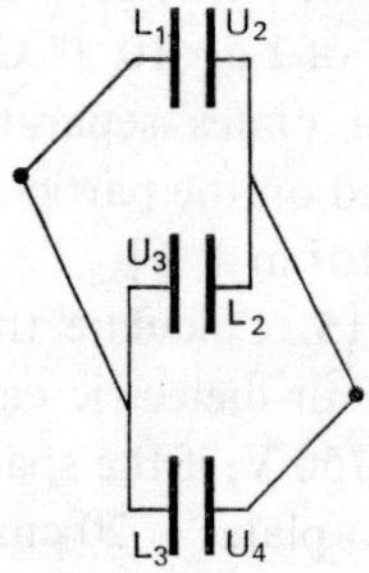

FIG. 13.5. Equivalent connections of multiplate capacitor (ex. 13.12).

But, for capacitors in parallel [see eqn. (12.5)], the total capacitance is the sum of the separate capacitances.

So for N plates, the total capacitance is

$$(N-1)\frac{\varepsilon A}{d}. \tag{13.13}$$

Solution

When $\varepsilon_r = 7$, $\varepsilon = \varepsilon_r \varepsilon_0 = 7 \times 8.85 \times 10^{-12}\ \text{Fm}^{-1}$.

Also $A = 2 \times 3 \times 10^{-4} = 6 \times 10^{-4}\ \text{m}^2$,

$d = 0.25 \times 10^{-3}$ m,

and $C = 0.0025 \times 10^{-6}$ F.

$\therefore$ from eqn. (13.13),

$$N = 1 + \frac{Cd}{\varepsilon A} = 1 + \frac{(0.0025 \times 10^{-6})(0.25 \times 10^{-3})}{(7 \times 8.85 \times 10^{-12})(6 \times 10^{-4})} = 1 + 16.8.$$

Therefore 18 plates are required.

13.13. A simple air-spaced capacitor having two plates separated by 3.0 mm holds a charge of 3.0×10^{-8} C at a p.d. of 360 V. Calculate the capacitance and the field strength in the air space. [83 pF; 1.2×10^{5} V m^{-1}.]

13.14. Particles of average mass 1.0×10^{-20} kg, each carrying a charge of 1.6×10^{-19} C, move freely in the air between two large parallel plates separated by 5.0 cm. Calculate the acceleration imposed on the particles when the p.d. between the plates is 500 V. [1.6×10^{5} m s^{-2}.]

13.15. Calculate the mean value of current in the leads of a simple air-dielectric capacitor connected to a battery of constant e.m.f. 150 V, if the spacing is halved in 1.0 ms. The area of each of the two plates is 20 cm^{2}, and the initial spacing is 2.0 mm. Neglect the resistance of the circuit. [1.3 μA.]

13.16. A multiplate capacitor having 15 plates each of area 3.0×4.0 cm has a capacitance of 0.0020 μF. Calculate the relative permittivity of the dielectric material if it is 0.50 mm thick. [6.7.]

13.17. A capacitor has 17 semicircular plates each of radius 3 cm, with 1 mm air spacing. Calculate the capacitance. [200 pF.]

13.18. A stack of 50 sheets of tin foil, each 10.0×5.0 cm, is interleaved with paper 0.10 mm thick having relative permittivity 3.0. Alternate sheets are connected together to form a capacitor to which a p.d. of 80 V is applied. Calculate (a) the electric field strength, (b) the electric flux density, (c) the charge on each sheet near the centre of the stack, and (d) the total capacitance. [(a) 0.8 MV m^{-1}; (b) 21 μC m^{-2}; (c) 2.1×10^{-7} C; (d) 6.5×10^{-8} F.]

Energy and Attractive Force

13.19. WORKED EXAMPLE

A simple parallel plate capacitor has two plates, each of area 20 cm^{2}, separated by 1.0 mm of air, and at a p.d. of 1000 V. Calculate (a) the force of the electric field on a dust particle having a charge of 4.8×10^{-19} C between the plates, and (b) the force of attraction of each plate toward the other.

Introduction

If one plate is fixed and the other is allowed to close up against it, with no change of charge Q, the work W done by the force of attraction F is given by $W = Fd$, where d is the spacing. F is independent of the spacing, since F depends on Q and E, both of which are constant.

Now the work comes from the stored energy which is given by $\frac{1}{2}QV$ [see eqn. (12.8)].

$$\therefore\ Fd = \tfrac{1}{2}QV$$

and
$$F = \frac{1}{2}Q\frac{V}{d} = \frac{1}{2}QE. \tag{13.14}$$

The factor $\frac{1}{2}$ in this expression corresponds with the division of the field into two components, one due to each plate, as in ex. 13.11, i.e. each plate has charge Q lying in the field of the other, and this field is $\frac{1}{2}E$, or half the total field between the plates.

Solution

(a) The force on the dust particle with charge q is due to the total field E of both plates.

Thus the force $= qE$ [see eqn. (13.2)]

$$= \frac{qV}{d} = 4.8 \times 10^{-19} \times \frac{1000}{10^{-3}} = 4.8 \times 10^{-13}\ \text{N}. \tag{a}$$

(b) The charge on each plate is given by $\pm Q = CV$, and $C = \frac{\varepsilon A}{d}$ [see eqn. (13.12)].

Then, from eqn. (13.14), the attractive force on each plate

$$= \frac{1}{2}QE = \frac{1}{2}\left(\frac{\varepsilon A}{d}V\right)\frac{V}{d} = \frac{1}{2}\varepsilon A\left(\frac{V}{d}\right)^2$$

$$= \frac{1}{2}(8.85\times10^{-12})(20\times10^{-4})\left(\frac{1000}{10^{-3}}\right)^2$$

$$= 8.9\times10^{-3}\ \text{N}. \qquad \text{(b)}$$

13.20. A simple capacitor of capacitance 10 pF is charged to a p.d. of 300 V. The two plates are each of area 10 cm² and are separated by a dielectric of relative permittivity 5.0. Calculate (a) the field strength, (b) the flux density, and (c) the force of attraction of one plate to the other. [(a) 68 kV/m; (b) 3 μC/m²; (c) 10^{-4} N.]

13.21. A simple capacitor has two plates, each of area 15 cm², separated by a sheet of dielectric 0.10 mm thick with relative permittivity 8.0. Calculate (a) the energy stored when the charge held is 1.0×10^{-8} C, and (b) the work required to remove the dielectric without altering the charge. [(a) 4.7×10^{-8} J; (b) 3.3×10^{-7} J.]

13.22. A battery of e.m.f. 100 V is connected across a simple capacitor consisting of two plates, each of area 250 cm², separated by air. Calculate the changes of charge and energy, when the spacing of the plates is increased from 1.0 mm to 3.0 mm. [1.5×10^{-8} C; 7.4×10^{-7} J.]

Spheres

13.23. Worked Example

Two concentric spheres have diameters of 10 cm and 12 cm respectively, and contain between them an oil of relative permittivity 3.0. Calculate (a) the energy stored when the potential difference between the spheres is 100 V and the outer sphere is connected to earthed surroundings, and (b) the charge held by the outer sphere when the inner sphere is uncharged and the outer sphere has a potential of 100 V relative to earthed surroundings.

Introduction

When a solid conducting sphere is charged negatively, the excess electrons distribute themselves around the surface only. Alternatively, when such a sphere is charged positively, the material of the interior retains its free electrons, while the surface layers have a deficiency and so display a positive charge. In both cases, positive and negative, the interior of the sphere remains uncharged. Also, it can be shown that a hollow conducting sphere has no interior field unless there is an enclosed charge, since at any interior point *A*, as in Fig. 13.6, the flux due to the charge on the smaller and closer portion *BCD* exactly cancels that of the charge on the larger but more distant portion *DEB*. Outside the sphere the surface charge gives a resultant flux which is radial and exactly as if all the charge were concentrated at the centre.

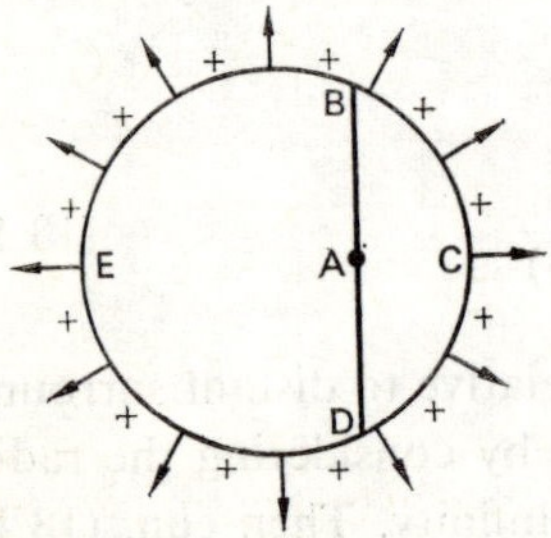

FIG. 13.6. A charged sphere (ex. 13.23).

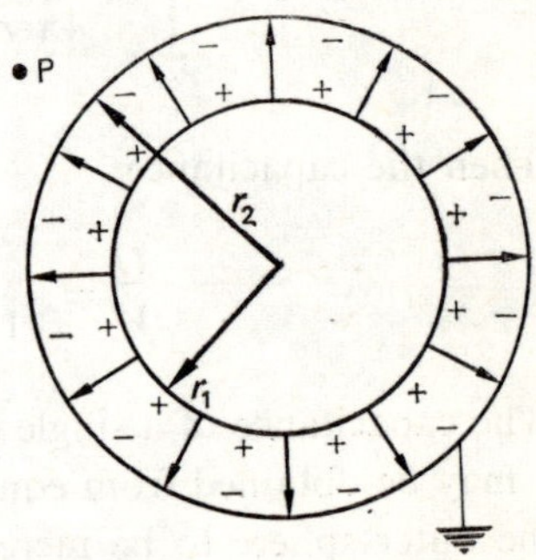

FIG. 13.7. Two concentric spheres (ex. 13.23).

When a second sphere is set concentrically around the first and is earthed, while the inner sphere is positively charged, as in Fig. 13.7, the positive charges attract electrons in the outer sphere.

This causes an electron flow up the earth lead, until the outer sphere gains a negative charge equal to the positive charge of the inner sphere. Then outside the outer sphere there is zero resultant flux, due to the charges shown, since at any point such as P both spheres act as if their charges were concentrated at the same centre.

Between the spheres the radial flux density and field strength are as for a point charge, i.e.

$$E = \frac{Q}{4\pi\varepsilon r^2}.$$

Now, from eqn. (13.8), with $x = r$,

$$\frac{dV}{dr} = -E = -\frac{Q}{4\pi\varepsilon r^2}.$$

Integrating between the limits set by the spheres of radii r_1 and r_2 gives the p.d. of the spheres

$$V = \int_{r_2}^{r_1} -\frac{Q\,dr}{4\pi\varepsilon r^2} = \frac{Q}{4\pi\varepsilon}\left(\frac{1}{r_1} - \frac{1}{r_2}\right).$$

Then the capacitance

$$\frac{Q}{V} = \frac{4\pi\varepsilon}{(1/r_1 - 1/r_2)}. \tag{13.15}$$

The capacitance of a single sphere relative to distant surroundings may be obtained from eqn. (13.15) by considering the radius of the outer sphere to be increased to infinity. Then eqn. (13.15) gives

$$C = \frac{4\pi\varepsilon}{\left(\frac{1}{r} - 0\right)} = 4\pi\varepsilon r, \tag{13.16}$$

where r is the radius of the single sphere.

Solution

(a) For the concentric spheres, $r_1 = 0.05$ m and $r_2 = 0.06$ m, $\varepsilon = 3.0 \times 8.85 \times 10^{-12}$ F m.

$$\text{Then from (13.15) the capacitance} = \frac{4\pi 3.0 \times 8.85 \times 10^{-12}}{\dfrac{1}{0.05} - \dfrac{1}{0.06}}$$

$$= 10^{-10} \text{ F}.$$

From eqn. (12.8) the energy stored when $V = 100$ V is given by

$$W = \tfrac{1}{2}CV^2 = \tfrac{1}{2} \times 10^{-10} \times (100)^2 = 5 \times 10^{-7} \text{ J}. \qquad \text{(a)}$$

(b) When the inner sphere is uncharged, and the outer is charged but not earthed, the flux radiates outward from the outer sphere alone. Then the capacitance of this sphere is given by eqn. (13.16) with $r = 0.06$ m and $\varepsilon = \varepsilon_0 = 8.85 \times 10^{-12}$ F m^{-1}.

Thus $C = 4\pi\varepsilon r = 4\pi \times 8.85 \times 10^{-12} \times 0.06 = 6.7 \times 10^{-12}$ F.

Then the charge held when $V = 100$ V is given by

$$CV = 6.7 \times 10^{-12} \times 100 = 6.7 \times 10^{-10} \text{ C}. \qquad \text{(b)}$$

13.24. An isolated sphere of diameter 10 cm carries a charge of 0.10 μC distributed uniformly over the surface in air. Calculate (a) the potential of the sphere, and (b) the electric flux density at a point 2.0 cm from the surface [(a) 18 kV; (b) 1.6×10^{-6} C m^2.]

13.25. Assuming that the insulation of air breaks down in a field of strength 3.0 kV mm^{-1}, calculate the maximum charge and energy which may be stored by an isolated sphere of diameter 10 cm. [8.4×10^{-7} C; 6.3×10^{-2} J.]

13.26. A single isolated sphere in air has a capacitance of 10 pF. Calculate the radius of a second sphere to be set concentrically around the first to increase the capacitance to 20 pF. [18 cm.]

13.27. Two thin concentric spheres have radii of 1.0 cm and 2.0 cm and charges of $+4.0 \times 10^{-9}$ C and -3.0×10^{-9} C respectively. Calculate the field strength and flux density (a) at a point midway between the spheres when the space contains a medium of relative permittivity 2.4, and (b) at a point in free space, 4.0 cm from the centre. [(a) 66 kV m^{-1}, 1.4 μC m^{-2}; (b) 5.6 kV m^{-1}, 0.05 μC m^{-2}.]

Measurement of Permittivity

13.28. Worked Example

An electrometer of negligible capacitance is connected to two parallel plates of a capacitor with 5.0 mm spacing, containing a sheet of dielectric material 2.0 mm thick. The capacitor is charged from a battery which is then disconnected. The electrometer deflection rises from 35 to 50 divisions when the material is removed. Calculate the relative permittivity of the material.

Introduction

Assuming the flux to be uniform between the plates, the density D is the same through both the material and the air gaps in series, i.e. $D = Q/A$ throughout, where $+Q$ and $-Q$ are the charges on the plates and A is the area of each face. However, the field strength in the air is different from that in the material if the permittivities are different (see Fig. 13.8).

Now the potential gradient in each case is uniform and numerically equal to the field strength. Thus, in the air,

$$\frac{V_0}{d_0} = \frac{D}{\varepsilon_0},$$

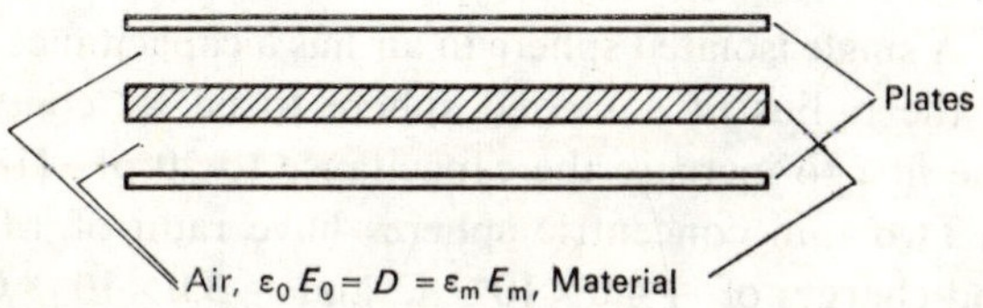

FIG. 13.8. Ex. 13.28.

where V_0 is the p.d. across the total width d_0 of air and ε_0 is the permittivity of the air (vacuum). Similarly, in the dielectric material m,

$$\frac{V_m}{d_m} = \frac{D}{\varepsilon_m}.$$

But the sum of the p.d.s in series equals the p.d. V between the plates.

$$V = V_0 + V_m = \frac{Dd_0}{\varepsilon_0} + \frac{Dd_m}{\varepsilon_m}$$

$$= D\left(\frac{d_0}{\varepsilon_0} + \frac{d_m}{\varepsilon_r \varepsilon_0}\right)$$

$$= \frac{Q}{\varepsilon_0 A}\left(d_0 + \frac{d_m}{\varepsilon_r}\right). \quad (13.17)$$

Solution

The electrometer deflections are proportional to the p.d.s with and without the material.

Thus with the material, $V = 35K$, where K is the proportionality constant of the electrometer; $d_m = 2\times 10^{-3}$ m; $d_0 = (5-2)\times 10^{-3}$ m.

Similarly, after removing the dielectric material, $V = 50K$; $d_m = 0$; $d_0 = 5\times 10^{-3}$ m.

Then, using eqn. (13.17) in each case,

$$\frac{35K}{50K} = \frac{\dfrac{Q}{\varepsilon_0 A}[(3\times 10^{-3}) + (2\times 10^{-3})/\varepsilon_r]}{\dfrac{Q}{\varepsilon_0 A}[(5\times 10^{-3}) + 0]}$$

Cancelling down, $$0.7 = \frac{3 + (2/\varepsilon_r)}{5}.$$

$$\therefore\ 2/\varepsilon_r = 3.5 - 3 = 0.5. \qquad \therefore\ \varepsilon_r = 4.$$

13.29. A simple air-spaced capacitor consists of two plates each of area 300 cm^2, spaced 1.0 mm apart. It is charged to a p.d. of 120 V and then discharged through a ballistic galvanometer which indicates the passage of a charge of 3.2×10^{-8} C. Calculate the permittivity of air. [8.9×10^{-12} F m^{-1}.]

13.30. A parallel plate capacitor with air dielectric is connected to an electrometer and charged from a battery which is then disconnected. A second capacitor of equal dimensions, but with mica dielectric, is then connected in parallel with the first. The electrometer deflection drops from 200 to 25. Calculate the relative permittivity of mica. [7.]

13.31. Two parallel metal plates are 1.8 mm apart in air. When a sheet of glass of thickness 1.8 mm is placed between the plates, their spacing must be increased to 3.3. mm to maintain constant capacitance. Calculate the relative permittivity of the glass. [6.]

13.32. Two parallel metal plates in air are separated by 4.0 mm, and have a capacitance of 25 pF. They are connected across an electrostatic voltmeter having a capacitance of 10 pF, and then charged to a p.d. of 1000 V by momentary connection to a high voltage d.c. supply. If then a sheet of dielectric material, 2.0 mm thick, is slid in between the plates, the p.d. indicated by the voltmeter drops to 750 V. Calculate the relative permittivity of the material. [2.7.]

CHAPTER 14

TRANSIENTS

Ballistic Galvanometer

14.1. Worked Example

A ballistic galvanometer is calibrated by discharging through it a capacitor of 0.20 μF, charged to a p.d. of 100 V. The first throw is 56 divisions and the next in the oscillation in the same direction is 46. The period is 5.0 s. Calculate the current sensitivity.

Introduction

A moving-coil galvanometer is ballistic if the inertia of the suspended coil assembly is large and the restoring torque and damping are small. Then if a pulse of charge is passed through the coil, the pulse may be completed before appreciable movement of the coil has occurred and the amplitude of the consequent oscillation is proportional to the total charge of the pulse.

Now from eqn. (1.7) the couple due to current i in the coil at any instant is given by

$$NABi, \tag{14.1}$$

where N is the number of turns of the coil, A is the area enclosed by each turn, and B is the flux density of the radial field in which the coil moves. If i flows for a short time δt, the coil is accelerated from rest to angular velocity ω_0, with negligible displacement of the coil, and the mechanical energy delivered to the coil

$$= \text{(couple)} \times \text{(mean angular velocity)} \times \text{(duration)}$$
$$= (NABi) \times (\omega_0/2) \times \delta t.$$

But the total charge of the pulse $Q = i\,\delta t$.

$$\therefore \text{ energy} = \tfrac{1}{2} NAB\omega_0 Q.$$

Now as the suspension becomes twisted, this energy is converted to elastic energy, given by (angular "throw")×(mean elastic couple)

$$\therefore \tfrac{1}{2} NAB\omega_0 Q = \Delta\theta \times \tfrac{1}{2} c\,\Delta\theta, \qquad (14.2)$$

where c is the torsion constant, i.e. the elastic couple per unit angle of twist, and $\Delta\theta$ is the angular throw.

Assuming the torsional oscillation of the coil to be simple harmonic motion, the maximum velocity = 2π (amplitude)/(period)

$$\therefore \omega_0 = 2\pi\,\Delta\theta/T, \qquad (14.3)$$

where T is the period.

Then substituting for ω_0 in eqn. (14.2),

$$\tfrac{1}{2} NAB(2\pi\,\Delta\theta/T)Q = \tfrac{1}{2} c\,\Delta\theta^2.$$

$$\therefore \frac{\Delta\theta}{Q} = \frac{2\pi}{T} \times \frac{NAB}{c}. \qquad (14.4)$$

Now for a steady current I, the steady deflection θ occurs when the couple $= c\theta = NABI$ from eqn. (14.1).

$$\therefore NAB/c = \theta/I. \qquad (14.5)$$

Then substituting in eqn. (14.4),

$$\frac{\Delta\theta}{Q} = \frac{2\pi}{T} \times \frac{\theta}{I}, \qquad (14.6)$$

where $\Delta\theta/Q$ is the "ideal charge sensitivity" of the galvanometer for a pulse of charge, θ/I is the "current sensitivity" for a steady current, and T is the period of oscillation. Note that eqn. (14.6) still holds if θ and $\Delta\theta$ are both measured in scale divisions rather than radians.

Also note that the reflected light of a mirror galvanometer is deflected through twice the angle turned by the mirror.

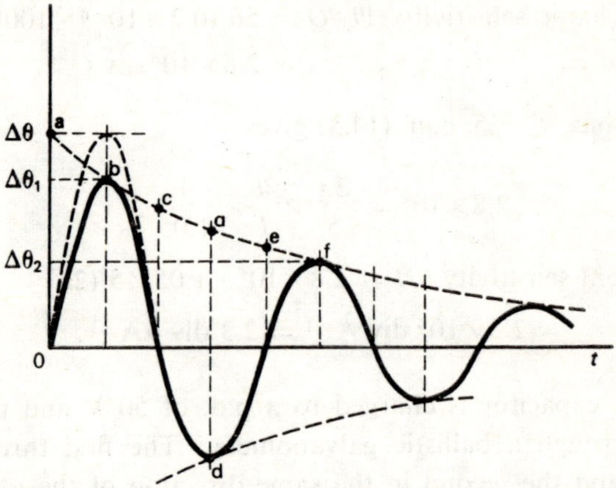

FIG. 14.1. Oscillatory deflection of a ballistic galvanometer (ex. 14.1).

In practice the damping is not entirely negligible, and the oscillation decreases as in Fig. 14.1. Thus the first real throw $\Delta\theta_1$ is less than the ideal $\Delta\theta$. The ordinates of the points *a*, *b*, *c*, etc., follow a geometric progression of common ratio *r*. The ratio of successive opposite deflections is r^2, and is known as the half period decrement.

So $\Delta\theta/\Delta\theta_1 = r$, and in Fig. 14.1,

$$\Delta\theta_1/\Delta\theta_2 = r^4. \tag{14.7}$$

Thus the real charge sensitivity

$$\frac{\Delta\theta_1}{Q} = \frac{2\pi}{T}\times\frac{\theta}{I}\times\frac{1}{r}. \tag{14.8}$$

Solution

In the question $\dfrac{\Delta\theta_1}{\Delta\theta_2} = \dfrac{56}{46} = 1.218 = r^4. \quad \therefore\ r = 1.05.$

The charge Q causing the ballistic deflection $= CV$, where C is the capacitance and V is the p.d.

$$\therefore \text{ real charge sensitivity } \Delta\theta_1/Q = 56/(0.2\times10^{-6}\times100)$$
$$= 2.8\times10^6 \text{ div C}^{-1}.$$

Then, since $T = 5$, eqn. (14.8) gives

$$2.8\times10^6 = \frac{2\pi}{5}\times\frac{\theta}{I}\times\frac{1}{1.05}\,.$$

$$\therefore \text{ current sensitivity } \theta/I = 2.8\times10^6\times1.05\times5/(2\pi)$$
$$= 2.3\times10^6 \text{ div A}^{-1} = 2.3 \text{ div } \mu\text{A}^{-1}.$$

14.2. A capacitor is charged to a p.d. of 50 V and then discharged through a ballistic galvanometer. The first throw is 55 divisions, and the second in the same direction of the oscillation is 42 after a time of 4.0 s. The galvanometer is known to give a steady deflection of 25 divisions for a steady current of 50 μA. Calculate the capacitance. [1.5 μF.]

14.3. A ballistic galvanometer has a rectangular coil of area 2.0 cm^2, wound with 150 turns. The coil swings with negligible damping and with a period of 5.0 s, in a radial field of 0.10 T. The suspension has a torsion constant of 2.0×10^{-7} Nm rad^{-1}. Calculate (a) the constant current for a mean deflection of 2.0 cm on a scale at a distance of 1.0 m from the galvanometer mirror, and (b) the charge sensitivity. [(a) $\frac{2}{3}$ μA; (b) 3.8 cm μC^{-1}.]

14.4. A long, air-cored solenoid, of diameter 5.0 cm, has 15 turns per cm of length and is surrounded at the centre by a secondary winding of 500 turns and 20 Ω resistance. The secondary winding is connected to a ballistic galvanometer of resistance 50 Ω. When a primary current of 2.0 A is suddenly reversed, the galvanometer throw is 15 div. Calculate the charge sensitivity of the galvanometer. [1.4×10^5 div C^{-1}.]

14.5. A ballistic galvanometer has a coil of 150 turns of wire, with effective length of 1.5 cm and width of 1.5 cm. The moment of inertia of the coil is 2.0×10^{-7} kg m^2, and the magnetic field has flux density of 0.40 T. Calculate (a) the torsion constant required for a current sensitivity of 1.0 deg μA^{-1}, (b) the period, and (c) the charge sensitivity. [(a) 7.7×10^{-7} Nm rad^{-1}; (b) 3.2 s; (c) 1.8 deg μC^{-1}.]

Discharge of Capacitor

14.6. Worked Example

A capacitor of 5.0 μF discharges through a resistor of 10 MΩ from an initial p.d. of 250 V. Calculate (a) the time required for the charge to decrease by 20%, (b) the p.d. when the discharge has run for a period equal to the time constant, and (c) the current when the discharge has run for a period equal to three times the time constant.

Introduction

When a capacitor discharges through a resistor, as in Fig. 14.2, they have the same p.d. v. Thus at any instant the capacitor charge q and the discharge current i are given by

$$v = q/C = iR. \tag{14.9}$$

Then v and i decrease in direct proportion to q, shown in Fig. 14.3, starting at time $t = 0$ from values given by

$$v_0 = q_0/C = i_0 R. \tag{14.10}$$

Now since the current is the rate of flow of charge at any time, the gradient $dq/dt = -i$. But from eqn. (14.9), $i = q/CR$

$$\therefore \; dq/dt = -q/CR. \tag{14.11}$$

Therefore, if at any time the rate of flow of q became constant, as shown by a tangent in Fig. 14.3, the discharge would be completed in a period of time equal to CR, called the time constant τ.

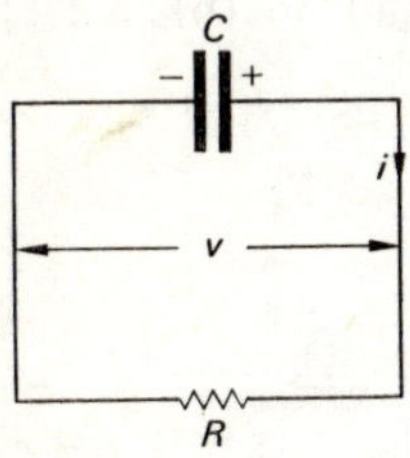

FIG. 14.2. Capacitor discharging through a resistor.

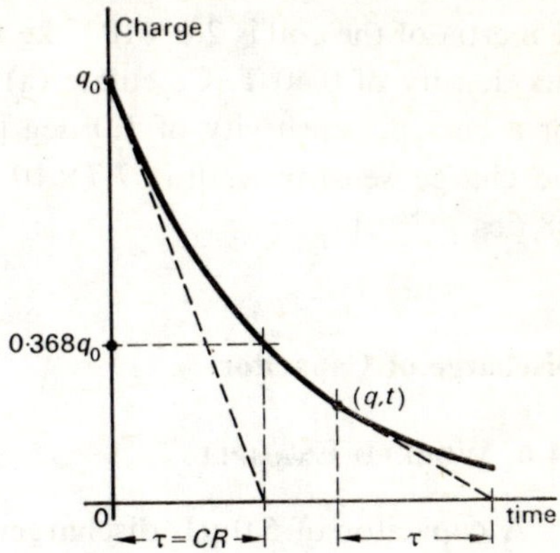

FIG. 14.3. Current of discharging capacitor (ex. 14.6).

The equation of the curve is the solution of eqn. (14.11), which with τ, becomes

$$dq/dt = -q/\tau. \tag{14.12}$$

Now, rearranging and integrating,

$$\int -\frac{dt}{\tau} = \int \frac{dq}{q}.$$

$$\therefore \quad -t/\tau = (\log_e q)+k,$$

where k is a constant, found by substituting $q = q_0$ when $t = 0$.

$$\therefore \; k = -\log_e q_0 .$$

$$\therefore \; -t/\tau = \log_e q - \log_e q_0 = \log_e (q/q_0).$$

$$\therefore \; e^{-t/\tau} = q/q_0 \quad \text{or} \quad q = q_0 e^{-t/\tau}. \tag{14.13}$$

Similarly, since $v = q/C$, $\quad v = v_0 e^{-t/\tau}$ (14.14)

Also, since $i = v/R$, $\quad i = i_0 e^{-t/\tau}$. (14.15)

In each of eqns. (14.13), (14.14), and (14.15), the instantaneous value, after a time t equal to τ, is given by the ratio

$$e^{-\tau/\tau} = e^{-1} = 0.368. \quad (14.16)$$

Solution

For (a), $q = q_0 - 20\%\ q_0 = 0.8q_0$.

Then eqn. (14.14) gives $0.8q_0 = q_0 e^{-t/\tau}$.

$$\therefore\ e^{t/\tau} = 1.25.$$

$$\therefore\ t/\tau = \log_e(1.25) = 0.223.$$

But $\tau = CR = 5\times10^{-6}\times10\times10^6 = 50$ s.

$$\therefore\ t = 0.223\times50 = 11 \text{ s.} \quad \text{(a)}$$

(b) From eqn. (14.16), with $v_0 = 250$, after a time $t = \tau$,

$$v = 250\times0.368 = 92 \text{ V.} \quad \text{(b)}$$

(c) When $t = 3\tau$, $e^{-t/\tau} = e^{-3} = 0.0498$.

But $i_0 = \dfrac{v_0}{R} = \dfrac{250}{10\times10^6} = 25\times10^{-6}$ A $= 25\ \mu$A.

$\therefore$ from eqn. (14.15), when $t = 3\tau$, $i = 25\times0.0498$

$= 1.2\ \mu$A. (c)

14.7. A capacitor discharges through a resistor of 4.0 MΩ, from an initial p.d. of 100 V, with a time constant of 2.0 s. Calculate (a) the capacitance, and (b) the p.d. at intervals of 1.0 s, throughout a period of 5 s from the start. [(a) 0.50 μF; (b) 61, 37, 22, 14, 8.2 V.]

14.8. The time taken by a capacitor to discharge by 50% in a non-inductive circuit is 1.0 s. Calculate the percentage of the initial charge remaining after a discharge period of (a) 2.0 s, (b) 0.50 s. Also calculate the time at which the remaining charge is (c) 6.25%, and (d) 10%. [(a) 25%; (b) 71%; (c) 4 s; (d) 3.3 s.]

14.9. The insulation resistance of a length of coaxial cable is tested by charging the conductors with the far end open-circuited, and observing the decay of charge caused by leakage through the insulation. The p.d. falls from 200 V to 160 V in 10 s. The capacitance is 200 pF. Calculate the insulation resistance. [2.2×10^{11} Ω.]

14.10. A capacitor of 10 μF is charged to a p.d. of 120 V and then isolated. The charge falls to 0.25 of its initial value in an interval of 2.0 min due to insulation leakage. Calculate (a) the insulation resistance, (b) the discharge current at the end of the interval, and (c) the time for the charge to reach 0.10 of the initial value. [(a) 8.7 MΩ; (b) 3.5 μA; (c) 200 s.]

Charging of Capacitor

14.11. Worked Example

A battery of e.m.f. 50 V is switched across an uncharged capacitor of 2.0 μF with a resistor of 0.25 MΩ in series. Calculate both the capacitor p.d. and the charging current at the following times after switching: 0, 0.25, 0.50, 1.0, 1.5, 2.0, 2.5 s.

Introduction

When a capacitor charges through a resistor, as in Fig. 14.4, the sum of their p.d.s equals the e.m.f. Therefore at any instant, the capacitor charge q and charging current i are given by

$$E = q/C + iR, \tag{14.17}$$

Then, as q increases, i decreases as shown in Fig. 14.5. At time $t = 0$, $q = 0$ and the initial current given by eqn. (14.17) is

$$i_0 = E/R. \tag{14.18}$$

The gradient of the curve at any point is found by differentiating eqn. (14.17) and substituting $i = dq/dt$.

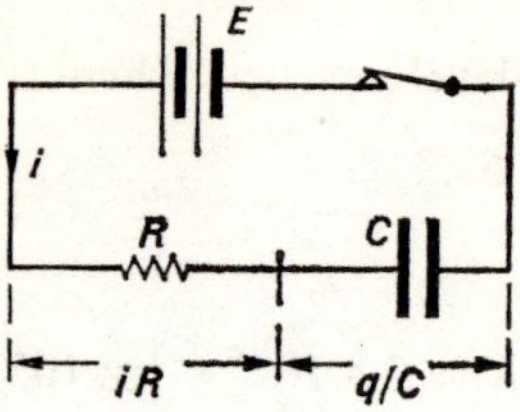

FIG. 14.4. Capacitor charging through resistor (ex. 14.11).

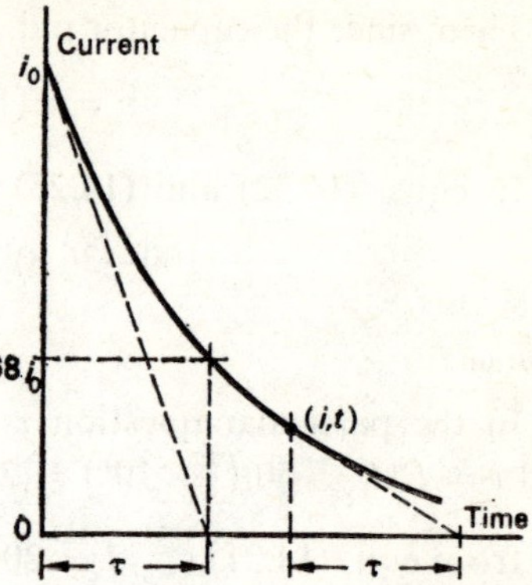

FIG. 14.5. Current of charging capacitor (ex. 14.11).

Thus $$0 = \frac{1}{C}\frac{dq}{dt} + \frac{1}{R}\frac{di}{dt}$$

and $$\frac{di}{dt} = -\frac{i}{CR}. \tag{14.19}$$

Therefore, if at any time the rate of fall of current were to become constant, the charging would be completed in the same constant time $\tau = CR$ as for discharge (ex. 14.6).

Then $$di/dt = -i/\tau. \tag{14.20}$$

This is the same form of equation as (14.12), and has the same form of solution.

Thus

$$i = i_0 e^{-t/\tau} \tag{14.21}$$

and $i = 0.368\ i_0$ when $t = \tau$.

The solution for q can be obtained from eqns. (14.17), (14.18), and (14.21):

$$q = EC - CRi \quad \text{from eqn. (14.17),}$$
$$= EC - CRi_0 e^{-t/\tau} \quad \text{using eqn. (14.21),}$$
$$= EC - ECe^{-t/\tau} \quad \text{using eqn. (14.18),}$$
$$= EC(1 - e^{-t/\tau}). \tag{14.22}$$

Then, since the capacitor p.d. $v = q/C$,

$$v = E(1 - e^{-t/\tau}). \qquad (14.23)$$

In eqns. (14.22) and (14.23), the level reached when $t = \tau$ is

$$(100 - 36.8)\% = 63.2\%.$$

Solution

In the particular question $\tau = CR = 2\times10^{-6}\times\frac{1}{4}\times10^{6} = \frac{1}{2}$ s and $i = E/R = 50/(\frac{1}{4}\times10^{6}) = 200$ μA.

$\therefore$ from eqn. (14.21), $\quad i = 200e^{-2t}$ μA,

and from eqn. (14.23), $\quad v = 50(1 - e^{-2t})$ V.

Figure 14.6 displays the values obtained from these equations, using reference tables of e^{-x}. Figure 14.7 shows the curves approximately.

t (s)	e^{-2t}	i (μA)	$1-e^{-2t}$	v (V)
0	1·0	200	0	0
0·25	0·607	120	0·393	20
0·50	0·368	74	0·632	32
1·0	0·135	27	0·865	43
1·5	0·050	10	0·950	48
2·0	0·018	3·7	0·982	49
2·5	0·0067	1·3	0·993	50

FIG. 14.6. Tabulated results for ex. 14.11.

14.12. An uncharged capacitor of 4.0 μF is connected in series with a resistor of 100 kΩ to a d.c. supply of 100 V. After allowing the capacitor to become fully charged, the supply is disconnected and the resistor is connected directly across the capacitor. Calculate the p.d. of the capacitor and the current through the resistor 1.0 s

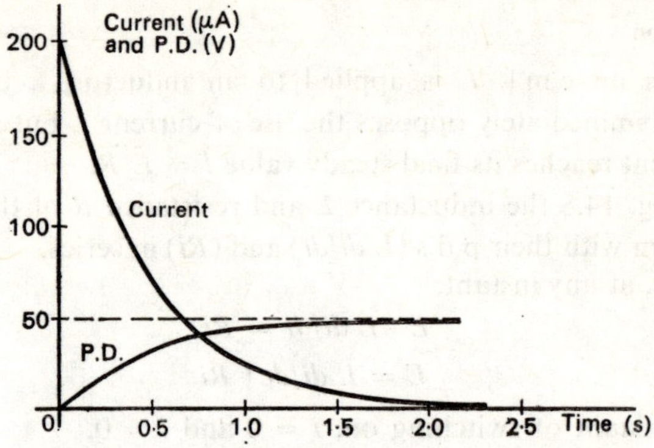

FIG. 14.7. Graphical results for ex. 14.11.

after the start of (a) the discharging current, and (b) the charging current. [(a) 8.2 V, 82 µA; (b) 92 V, 82 µA.]

14.13. A neon tube whose striking voltage is 180 V is connected in parallel with a capacitor of 8.0 µF. A battery of e.m.f. 240 V is then connected in series with a resistor to the capacitor terminals. Calculate (a) the resistance which delays the striking for 5.0 s after switching on, and (b) the time delay if the resistance is 1.0 MΩ. [(a) 450 kΩ; (b) 11 s.]

Growth of Current in Inductor

14.14. WORKED EXAMPLE

A coil of inductance 0.10 H and resistance 20 Ω is connected directly to a supply of e.m.f. 10 V. Calculate (a) the initial rate of increase of current, (b) the time constant, (c) the current after a time equal to half the time constant, and (d) the time taken for the current to reach 250 mA.

Introduction

When an e.m.f. E is applied to an inductor, a back e.m.f. ($L\,di/dt$) immediately opposes the rise of current i, but dies out as the current reaches its final steady value $I = E/R$.

In Fig. 14.8 the inductance L and resistance R of the inductor are shown with their p.d.s ($L\,di/dt$) and (Ri) in series.

Thus, at any instant,

$$E - L\,di/dt = Ri$$

or

$$E = L\,di/dt + Ri. \tag{14.24}$$

At the instant of switching on, $t = 0$ and $i = 0$.

Therefore, from eqn. (14.24), $\quad (di/dt)_0 = E/L, \qquad (14.25)$

giving the gradient of the i/t graph at the origin (Fig. 14.9).

If this initial rate of growth were to continue, i would reach its final value I in a time τ called the time constant. Thus $(di/dt)_0 = I/\tau$

$$\therefore\ \tau = I/(di/dt)_0 = E/R \div E/L = L/R. \tag{14.26}$$

The equation of the curve is the solution of eqn. (14.24), which

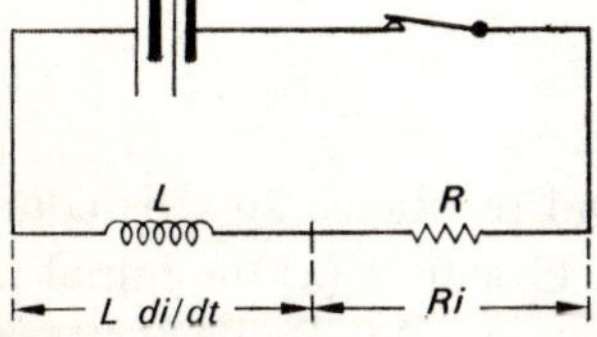

FIG. 14.8. Series R, L circuit (ex. 14.14).

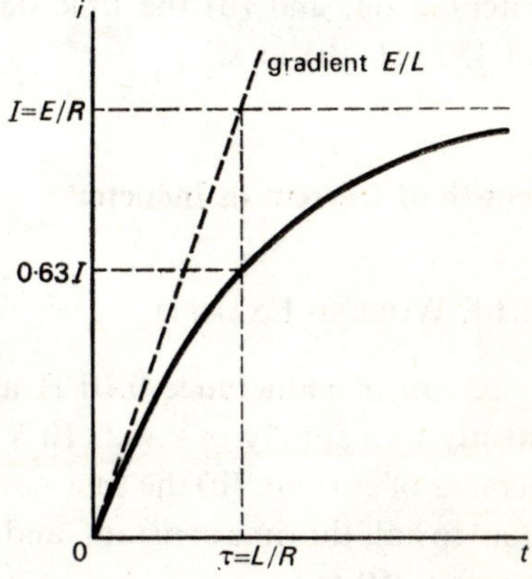

FIG. 14.9. Current growth in a series R, L circuit (ex. 14.14).

may be found by substituting $E/R = I$ and $L/R = \tau$, and then rearranging and integrating. Thus $I = i+\tau\, di/dt$.

$$\therefore \int \frac{dt}{\tau} = \int \frac{di}{I-i}. \qquad \therefore t/\tau = -\log_e (I-i)+k,$$

where k is a constant found by substituting $i = 0$ when $t = 0$.

$$\therefore k = \log_e I. \qquad \therefore -t/\tau = \log_e (I-i)-\log_e I$$

$$= \log_e (I-i)/I. \qquad (14.27)$$

$$\therefore e^{-t/\tau} = (I-i)/I. \qquad \therefore i = I(1-e^{-t/\tau}). \qquad (14.28)$$

Then when $t = \tau$, $i = I(1-e^{-\tau/\tau}) = I(1-e^{-1})$.
But $e^{-1} = 0.368$. Therefore as shown in Fig. 14.9,

$$i_\tau = 0.63I. \qquad (14.29)$$

Solution

(a) From eqn. (14.25),

$$(di/dt)_0 = E/L = 10/0.1 = 100 \text{ A s}^{-1}. \qquad \text{(a)}$$

(b) From eqn. (14.26),

$$\tau = L/R = 0.1/20 = 5\times10^{-3} \text{ s} = 5 \text{ ms}. \qquad \text{(b)}$$

(c) When $t = \tau/2$, with $I = E/R = 10/20 = 0.5$ A, eqn. (14.28) gives

$$i = 0.5(1-e^{-1/2}) = 0.5(1-0.607)$$

$$= 0.20 \text{ A}. \qquad \text{(c)}$$

(d) When $i = 0.250$ A, with $\tau = 5\times10^{-3}$ s, eqn. (14.27) gives

$$-t/(5\times10^{-3}) = \log_e (0.5-0.25)/0.5$$

$$= \log_e \tfrac{1}{2} = -0.693.$$

$$\therefore t = 0.693\times5\times10^{-3} \text{ s} = 3.5 \text{ ms}. \qquad \text{(d)}$$

14.15. A supply of e.m.f. 10 V is applied directly across the terminals of an inductor of 10 Ω resistance and 0.10 H inductance.

Calculate the current after time intervals equal to the following multiples of the time constant: $\frac{1}{2}$, 1, 2, 3, 4, 5. [0.39, 0.63, 0.86, 0.95, 0.98, 0.99 A.]

14.16. An inductor has a coil of 1000 turns with resistance of 25 Ω. A current of 5.0 A produces a flux of 10 mWb in the core. Calculate (a) the time constant, and (b) the initial rate of rise of current when an e.m.f. of 30 V is suddenly applied to the coil. [(a) 0.08 s; (b) 15 A s^{-1}.]

14.17. A coil of resistance 10 Ω and self-inductance 1.0 H has a constant direct e.m.f. applied to its terminals. At some instant, the current is 1.5 A and is rising at the rate of 5.0 A s^{-1}. Calculate (a) the e.m.f., (b) the time constant, (c) the final current, (d) the initial rate of rise of current, and (e) the energy stored and power dissipated at the given instant. [(a) 20 V; (b) 0.1 s; (c) 2 A; (d) 20 A s^{-1}, (e) 1.1 J; 23 W.]

14.18. A relay of resistance 10 Ω and inductance 5.0 H is operated on a supply of e.m.f. (a) 2.0 V, and (b) 2.5 V. Calculate the time lag in its operation after switching on if the operating current is 150 mA. [(a) 0.69 s; (b) 0.46 s.]

CHAPTER 15

ALTERNATING E.M.F., P.D., AND CURRENT

Simple Alternator

15.1. WORKED EXAMPLE

A simple alternator has a rectangular coil of 20 turns, 6.0×8.0 cm, pivoted about an axis passing through the centres of the 6 cm sides. The coil rotates at a rate of 3000 rev min^{-1} in a uniform magnetic field of 0.20 T perpendicular to the axis. Calculate (a) the frequency of the e.m.f. generated, (b) its period, (c) its maximum value, and (d) the e.m.f. at an instant 0.0030 s after zero e.m.f.

Introduction

Figure 15.1 shows the coil *ABCD* rotating between flat-faced pole pieces N and S which provide vertical magnetic flux which is assumed to be uniform. As the coil rotates e.m.f.s are generated in the sides *AB* and *CD* which cut across the flux, but not in *AD* or *BC* which never cut across it. When one e.m.f. is from *A* to *B* the other is from *C* to *D*, since the movements of *AB* and *CD* are always opposite in direction relative to the flux. Thus the e.m.f.s are in series round the coil, and may be led out through slip rings and brushes.

The total e.m.f. at any instant depends upon the effective velocity with which *AB* and *CD* cut across the flux. Thus if v is the velocity

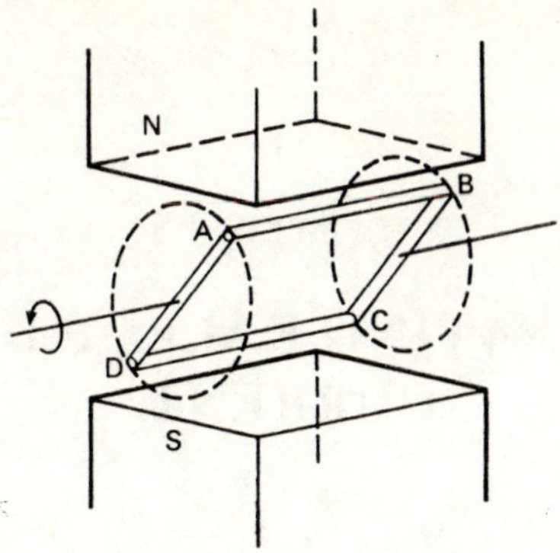

FIG. 15.1. Coil rotating in a magnetic field.

of *AB* or *CD* in the circular path of Figs. 15.2 (and 15.1), the component of velocity perpendicular to the flux at time t is $v \sin \theta$, when the coil makes an angle θ with the cross-field position. An expression for the e.m.f. is then obtained by modifying eqn. (8.2).

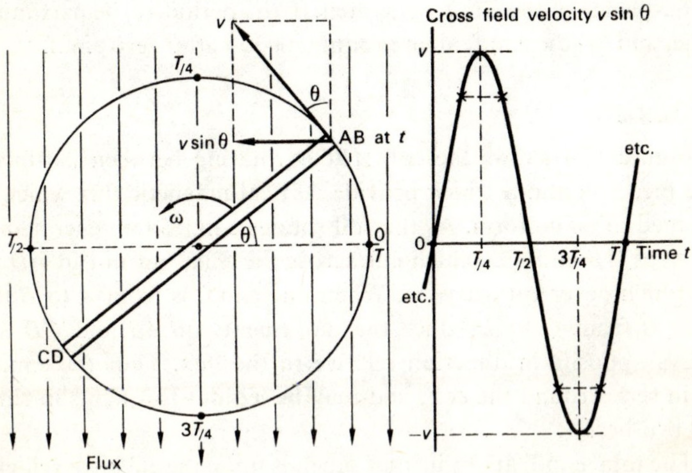

FIG. 15.2. Sinusoidal e.m.f. induced in a coil (ex. 15.1).

Thus for a coil of N turns, with $AB = CD = l$, the total active length $= 2\ Nl$, and eqn. (8.2) gives e.m.f.

$$e = Bv \sin \theta\ 2Nl, \tag{15.1}$$

where B is the flux density.

Now if f is the frequency of rotation in rev s^{-1}, and $2\pi r$ is the circumference of the circle, $v = 2\pi rf$.

Therefore from eqn. (15.1)

$$e = B2\pi rf \sin \theta\ 2Nl. \tag{15.2}$$

But $2lr$ = area A enclosed by each turn of the coil. Also, since the angle turned in each revolution is 2π rad,

$$2\pi f = \text{angular velocity } \omega \text{ in rad s}^{-1}. \tag{15.3}$$

Therefore from (15.2) $\quad e = BAN\omega \sin \theta. \quad$ (15.4)

The maximum e.m.f. E_m occurs when $\theta = \pi/2$, since the velocities of AB and CD are then perpendicular to the flux.

So, from eqn. (15.4), $\quad E_m = BAN\omega \quad$ (15.5)

and at any angle $\theta \quad e = E_m \sin \theta, \quad$ (15.6)

where, if t is the time in seconds for the coil to turn through the angle θ,

$$\theta = \omega t. \tag{15.7}$$

The right-hand portion of Fig. 15.2 shows the variation of $v \sin \theta$ with t, and therefore, from eqn. (15.1), shows the form of variation of e.m.f. also, one cycle being completed in one revolution of periodic time $T = 1/f$. Thus f is the frequency of both the rotation and the e.m.f.

Solution

(a) For the stated rate of rotation of 3000 rev min^{-1},

$$f = 3000/60 \text{ cycles/second} = 50 \text{ hertz (Hz)}. \tag{a}$$

(b) Periodic time or period $T = 1/50$ s $= 20$ ms. $\quad$ (b)

(c) Equation (15.5) gives $E_m = BAN\omega$, where A is in m^2 and ω in rad s^{-1}.

$$\therefore\ E_m = 0.2\times(6\times8\times10^{-4})\times20\times(2\pi\times50) = 6.0\ \text{V}. \qquad \text{(c)}$$

(d) Zero e.m.f. occurs when $t = 0$ or $T/2$. Then, 0.003 s after the instant $t = 0$, eqn. (15.7) gives

$$\theta = (2\pi\times50)\times0.003 = 0.3\pi\ \text{rad} = 0.3\pi\times180/\pi\ \text{degrees} = 54°,$$

as shown approximately in Fig. 15.2. At this phase, eqn. (15.6) gives

$$e = 6.0\times\sin 54° = 6.0\times0.809 = 4.9\ \text{V}. \qquad \text{(d)}$$

This value occurs twice in each direction in each cycle (see X in Fig. 15.2).

15.2. A simple alternator has a single flat coil of 30 turns, 4.0×3.0 cm, pivoted about an axis through the centres of the 3 cm sides. The coil turns at a rate of 6000 rev min^{-1} in a uniform magnetic field perpendicular to the axis. The maximum value of the e.m.f. generated is 6.8 V. Calculate (a) the flux density, (b) the time between successive instants of zero e.m.f., and (c) the e.m.f. at an instant 1.5 ms after zero e.m.f. [(a) 0.3 T; (b) 5 ms; (c) 5.5 V.]

15.3. A rectangular coil of 50 turns, 5.0×8.0 cm, is pivoted as in a simple alternator. Calculate the rate at which the coil must turn in a field of flux density 80 mT to generate an e.m.f. of 10 V peak value. [6000 rev min^{-1}.]

15.4. The maximum e.m.f. of an alternator is 10 V at a frequency of 50 Hz. Calculate (a) the e.m.f.s at instants 2.0 ms and 6.0 ms after zero e.m.f., and (b) the time for the e.m.f. to change from 0 to 10 V, and 0 to 9.0 V. [(a) 5.9 V, 9.5 V; (b) 5.0 ms, 3.6 ms.]

Phasors and Instantaneous Values

15.5. Worked Example

A sinusoidal current has an instantaneous value of 10 mA a fifth of a cycle after passing through zero. Calculate (a) the maximum value, and (b) the instantaneous value three-fifths of a cycle after zero.

Introduction

If a sinusoidal e.m.f. is applied to a circuit which draws a current proportional to the e.m.f. at every instant (Ohm's law), then the current is given by an equation similar to that of the e.m.f.

Thus, from eqns. (15.6) and (15.7),

$$e = E_m \sin \omega t \tag{15.8}$$

and now

$$i = I_m \sin \omega t, \tag{15.9}$$

where i is the instantaneous current at time t after a zero value, I_m is the maximum or peak value or "amplitude" of current, ω is the angular velocity of simple alternator in rad s^{-1}, and also $\omega = 2\pi f$, where f is frequency in hertz, Hz, or cycles per second.

For any sinusoidally varying quantity such as e or i, the amplitude or maximum value may be represented to scale by a rotating radius of an "auxiliary" circle (as in Fig. 15.3). As the radius CA rotates at constant angular velocity ω, A rises and falls sinusoidally above and below the axis, since $AB = CA \sin \omega t$. At the instant shown t is the time from the instant when A was at 0, with $t = 0$ and $AB = 0$. CA may be compared with the rotating coil of the alternator for which eqn. (15.2) gives $e \propto r \sin \theta$.

The radius used in this way is called a phasor (or vector): by its angle it shows the phase or stage of the variation reached at the particular time; by its length it shows the amplitude; and by the

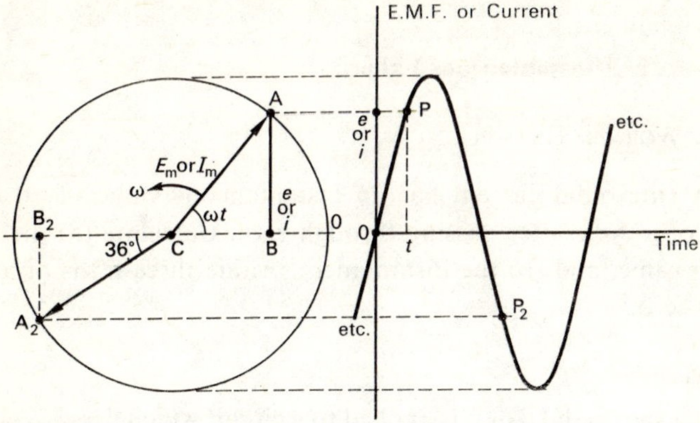

FIG. 15.3. EX. 15.5.

height of its tip it shows the instantaneous value. The instantaneous value is shown also by the sine wave of Fig. 15.3, where P is always at the same height as A.

Solution

(a) If the angle ωt is one-fifth of a cycle,

$$\omega t = 1/5 \times 2\pi \text{ rad} = 360°/5 = 72°.$$

Then if AB represents $i = 10$ mA, in Fig. 15.3, or from eqn. (15.9),

$$I_m = 10/\sin 72° = 10/0.951 = 10.5 \text{ mA.} \quad \text{(a)}$$

(b) When $t = 3/5 \times 2\pi$ rad $= 216°$, eqn. (15.9) gives

$$i = 10.5 \sin 216° = -10.5 \sin 36°$$

$$= -10.5 \times 0.588 = -6.2 \text{ mA.} \quad \text{(b)}$$

The points A_2 and P_2 in Fig. 15.3 refer to this instant, when $A_2B_2 = -CA_2 \sin 36°$, and the negative sign indicates that A_2 is below B_2. Currents (a) and (b) are in opposite directions.

15.6. A resistor of 24 Ω has a sinusoidal p.d. of 24 V at a frequency of 50 Hz. Calculate the instantaneous values of p.d. and current (a) 20 ms after zero p.d., and (b) 2.0 ms after zero p.d. [(a) 0, 0; (b) 14 V, 0.59 A.]

15.7. A sinusoidal alternating current has a peak value of 2.0 A and a frequency of 50 Hz. Calculate (a) the current 1.0 ms after peak, (b) the current 1.0 ms after zero, and (c) the times at which the current is half its peak value during one cycle. [(a) 1.9 A; (b) 0.62 A; (c) 1.7, 8.3, 12, 18 ms.]

15.8. A simple alternator has a coil of area 10 cm^2 wound with 25 turns of wire and pivoted to turn in a uniform magnetic field of flux density 0.75 T. The coil has a resistance of 10 Ω and is connected across an external resistor of 20 Ω. Calculate for a speed of 3000 rev min^{-1} (a) the maximum instantaneous current, and (b) the instantaneous current 4.0 ms after the maximum occurs. [(a) 0.20 A; (b) 0.061 A.]

15.9. A sinusoidal current is defined by the following values: $i = 0$, $t = 0$; $i = 10$ mA, $t = 2.0$ ms; $i = 11$ mA, $t = 4.0$ ms. Calculate the peak value and the frequency. [12 mA; 79 Hz.]

Power and R.M.S. Values

15.10. Worked Example

A current of maximum value 14.0 A is drawn by a pure resistor connected across a supply of sinusoidal p.d. 240 V. Calculate (a) the mean power, (b) the peak power, and (c) the resistance.

Introduction

The power, or rate at which the resistor turns electrical energy into heat, is $p = Ri^2$ at any instant at which the current is i and where R is the resistance. Since i varies during the cycle, p varies also.

$$\text{The mean value of } p = \text{mean value of } Ri^2$$
$$= R\times(\text{mean value of } i^2). \tag{15.10}$$

Now if $i = I_m \sin \omega t$, as in eqn. (15.9),

$$i^2 = I_m^2(\sin \omega t)^2 = I_m^2(\tfrac{1}{2} - \tfrac{1}{2} \cos 2\omega t). \qquad (15.11)$$

But the mean value of the term $\cos 2\omega t$, over any whole number of periods, is zero, as for any sinusoidal quantity (not squared).

$\therefore$ from eqn. (15.11) the mean value of $i^2 = I_m^2/2$. (15.12)

Equations (15.11) and (15.12) are illustrated by Fig. 15.4, where the graph of i^2 shows the sinusoidal shape of $\sin^2 \omega t$. The frequency of i^2 is twice that of i, since $(-i)^2 = +i^2$, in accordance with the fact that the power remains positive when the current reverses.

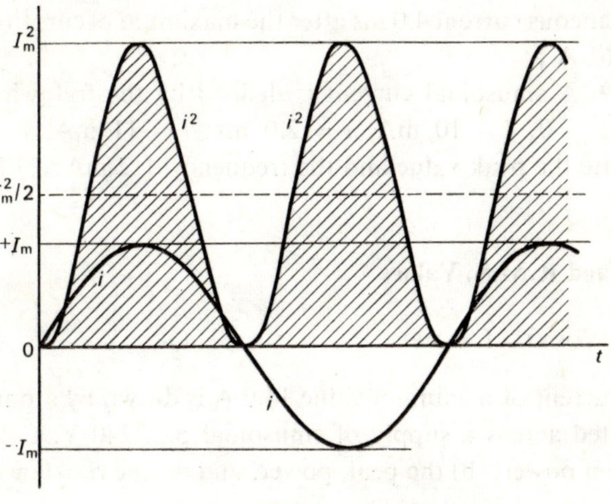

FIG. 15.4. Ex. 15.10.

Since the unshaded areas below the broken line equal the shaded areas above it, this line shows the mean level of i^2, as in eqn. (15.12). It also shows the constant value of I^2 required for equivalent power supplied by a direct current I.

Thus, for equivalent powers on a.c. and d.c.,

$$P = RI_m^2/2 = RI^2. \qquad (15.13)$$

Therefore the d.c. $I = \sqrt{(I_m^2/2)}$, called the root mean square value (r.m.s.) of the a.c.

$$\therefore\ I_{rms} = I_m/\sqrt{2} = 0.707I_m \tag{15.14}$$

and

$$P = RI_{rms}^2. \tag{15.15}$$

Similarly, if the sinusoidal p.d. $v = V_m \sin \omega t$, then, for a pure resistor (having no inductance or capacitance), the instantaneous power $p = v^2/R$, and the mean power

$$P = (V_m^2/2)R = V_{rms}^2/R. \tag{15.16}$$

Since $V_{rms} = I_{rms}R$, as on d.c., eqn. (15.15) or (15.16) gives P, for a pure resistor,

$$= V_{rms}I_{rms}. \tag{15.17}$$

An alternating e.m.f., p.d., or current is usually assumed to be of sinusoidal form, and the values quoted for E, V, or I are assumed to be r.m.s. unless otherwise stated.

Solution

(a) For $I_m = 14$ A, eqn. (15.14) gives $I_{rms} = 14/\sqrt{2}$

$$= 0.707\times 14$$

$$= 9.9 \text{ A}.$$

Then, assuming the stated value of 240 V to be the r.m.s. value, eqn. (15.17) gives mean power

$$P = 240\times 9.9 = 2380 \text{ W} = 2.38 \text{ kW}. \tag{a}$$

(b) The maximum or peak power occurs at instants when $i = \pm I_m$ and $v = \pm V_m$. Thus, from eqn. (15.13) or (15.16), and using answer (a),

$$\text{peak power} = RI_m^2 = V_m^2/R = 2P = 4.76 \text{ kW}. \tag{b}$$

Alternatively, maximum $p = V_mI_m = (240\times\sqrt{2})\times 14$

$$= 4760 \text{ W}. \tag{b}$$

(c) Since the r.m.s. values, in the case of a pure resistor, equal the equivalent values in a d.c. circuit,

$$R = V_{rms}/I_{rms} = 240/9.9 = 24.2\,\Omega. \tag{c}$$

Alternatively, for a pure resistor, R is given by eqn. (15.15) or (15.16) or by the ratio of V_m and I_m or of any corresponding values of v and i.

15.11. Calculate the peak sinusoidal values of (a) an alternating e.m.f. of 240 V, and (b) an alternating current of 13.0 A. [(a) 340 V; (b) 18.4 A.]

15.12. Calculate the resistance of a pure resistor which dissipates a mean power of 1.0 kW on an a.c. supply of peak sinusoidal e.m.f. 25 V. [0.31 Ω.]

15.13. Calculate the r.m.s. p.d. and mean power for a pure resistor of 50.0 Ω carrying a sinusoidal alternating current of maximum value 5.00 A. [177 V; 625 W.]

15.14. A pure resistor dissipates a peak power of 200 W when its sinusoidal alternating p.d. is 240 V. Calculate (a) the resistance, and (b) the peak current. [(a) 576 Ω; (b) 0.59 A.]

15.15. A current $i = 2.0 \sin(100\pi t)$ A flows in a pure resistor of 20 Ω. Calculate (a) the mean power, and (b) the direct p.d. needed for the same power. [(a) 40 W; (b) 28 V.]

15.16. The p.d. across a pure resistor of 30.0 Ω is given by $v = 150 \sin(700\pi t)$ volts. Calculate (a) the frequency, (b) the r.m.s. p.d., (c) the r.m.s. current, (d) the peak current, and (e) the mean power. [(a) 350 Hz; (b) 106 V; (c) 3.5 A; (d) 5.0 A; (e) 375 W.]

15.17. A flat coil has 30 turns of area 10 cm^2 and resistance of 25 Ω. It is rotated at 1000 rev min^{-1} about an axis which halves the area and lies perpendicular to a field of 0.50 T. When the coil is short circuited, find (a) the current flowing, and (b) the heat produced in the coil in 10 min. [(a) 44 mA r.m.s. at 17 Hz; (b) 30 J.]

Mean Values

15.18. Worked Example

A resistor of 100 Ω dissipates heat of 48×10^3 J in a time of 2.0 min when passing a fully rectified sinusoidal current. Calculate (a) the mean value of a positive half of the current wave, and (b) the total charge passed in the 2 minutes.

Introduction

The average or mean value of one complete cycle of a sinusoidal wave is zero, since the positive and negative pulses or halves of the cycle are exactly equal and opposite. If, however, the negative pulses are reversed to become positive, the wave is said to be fully rectified, and the mean value depends upon positive pulses only.

For this purpose, a simple alternator, as in ex. 15.1, may be modified by adding a commutator, which is a synchronous reversing switch, fitted to the rotor so as to operate each time the e.m.f. passes through zero.

To find and expression for the mean e.m.f. note that, between zeros, the rotating coil of Fig. 15.1 turns over, and the flux linkage is reversed from BAN to $-BAN$, where B is the flux density and N is the number of turns, each of area A. Also the time taken for this change is half a period $= (2\pi/\omega)/2$, where ω is the angular velocity.

Then the mean e.m.f. of a positive half wave

$$= \text{mean rate of change of flux linkage}$$

$$= \frac{BAN-(-BAN)}{(2\pi/\omega)/2} = \frac{2}{\pi}(BAN\omega)$$

$$= \frac{2}{\pi} \times \text{maximum e.m.f.} = 0.637E_m \qquad (15.18)$$

using eqn. (15.5).

The same relationship applies to alternating currents, since the factor $2/\pi$ depends only on the waveform being sinusoidal. Thus the mean current of a positive half wave

$$= 0.637 I_m. \qquad (15.19)$$

Alternatively, $\sin\theta$ may be integrated over a positive half cycle and divided by the corresponding change of θ, giving the mean of the positive half wave

$$= \frac{1}{\pi}\int_0^\pi \sin\theta \, d\theta = \frac{1}{\pi}\Big[-\cos\theta\Big]_0^\pi = \frac{1}{\pi}(1+1) = \frac{2}{\pi}.$$

Now in a fully rectified sinusoidal wave of current there are no intervals between the half waves and all have the same mean value I_{mean}. This value equals that of a direct current which would convey the same charge Q in the same time t.

Thus total charge passed,

$$Q = I_{\text{mean}} t. \qquad (15.20)$$

Solution

(a) The r.m.s. current is found from the power $P = I^2R = W/t$, where $R = 100\ \Omega$, energy $W = 48\times 10^3$ J, and time $t = 120$ s.

Thus $P = 48\times 10^3/120 = 400$ W,

and $I = \sqrt{(P/R)} = \sqrt{(400/100)} = 2$ A.

Now this r.m.s. value $I = I_m/\sqrt{2}$, from eqn. (15.14).

$\therefore$ the maximum value $I_m = 2\sqrt{2} = 2.83$ A.

Then, from eqn. (15.19), mean current of positive half wave

$$= 0.637\times 2.83$$

$$= 1.8 \text{ A.} \qquad \text{(a)}$$

(b) Using (a) eqn. (15.20) gives $Q = 1.8\times 120$

$$= 220 \text{ C.} \qquad \text{(b)}$$

15.19. Each pulse of the e.m.f. of a simple alternator has a mean value of 96 mV. The rotating coil is rectangular, with 30 turns, each 2.0×4.0 cm, and rotates at a constant rate of 50 rev s^{-1}, about an axis through the centres of the short sides and perpendicular to a uniform magnetic field. Calculate the flux density of the field. [20 mT.]

15.20. A flat square coil of 20 turns, 2.0×2.0 cm, is pivoted about an axis bisecting one pair of opposite sides and running perpendicular to a field of flux density 0.10 T. Calculate the frequency of rotation required to generate an e.m.f. of (a) half wave mean value 1.0 V, and (b) maximum value 1.0 V. [(a) 310 Hz; (b) 200 Hz.]

15.21. What is the mean value of sinusoidal a.c. giving the same mean power in a resistor as 10 A d.c.? [9 A.]

15.22. Calculate (a) the heat produced, and (b) the total charge passed in a resistor of 20 Ω running for 10 min on a fully rectified sinusoidal supply of peak e.m.f. 150 V. [(a) 34×10^4 J; (b) 2.9 kC.]

15.23. A straight piece of wire, 1.5 cm long, vibrates sinusoidally from side to side in a direction perpendicular to a magnetic field of 0.50 T. The vibration has an amplitude of 0.10 mm and a frequency of 1.0 kHz. Calculate the mean e.m.f. generated in a half cycle of one direction. [3 mV.]

Ideal Transformers

15.24. Worked Example

When a supply of 240 V, 50.0 Hz is connected to a certain transformer, the peak flux in the core is 1.10 mWb and the secondary e.m.f. is 500 V. The leakage flux, reluctance, and heat losses are assumed to be negligible. Calculate (a) the number of turns of the windings, (b) the currents in the windings when the output power is 200 W in a pure load resistor, and (c) the actual load resistance and the apparent primary resistance due to the load resistor of (b).

Introduction

Since the leakage flux is negligible, eqns. (11.3) and (11.4) apply. Thus in either winding the e.m.f.,

$$e = N\,d\varphi/dt, \qquad (15.21)$$

where N is the number of turns of the particular winding and $d\varphi/dt$ is the rate of change of flux linking with the winding.

Now if the e.m.f. is sinusoidal, $e = E_m \sin \omega t$, as in eqn. (15.8).

Then, with eqn. (15.21), $d\varphi/dt = e/N = (E_m/N) \sin \omega t$.

$$\therefore\ \varphi = \int (E_m/N) \sin \omega t\, dt$$
$$= -(E_m/\omega N) \cos \omega t + \text{(possible constant)}.$$

Therefore the alternating flux has an amplitude of

$$\Phi_m = E_m/\omega N$$

or

$$E_m = \Phi_m N\omega = B_m A N\omega, \qquad (15.22)$$

where B_m is the maximum flux density and A is the cross-section area.

Note that eqn. (15.22) applies also to a simple alternator, as in ex. 15.1, where the e.m.f. could be calculated by considering the variation of flux linkage as the coil rotates.

Now if the reluctance is negligible, so is the m.m.f. required to magnetize the core. Therefore the primary and secondary m.m.f.s are practically equal and opposite: any change in one without a corresponding change in the other would cause a large change in flux and induce e.m.f.s and currents which would restore the equality.

Thus $N_1 i_1 = N_2 i_2$ at every instant, and therefore the transformer ratio applies to currents as well as to e.m.f.s, as in ex. 11.7. Thus

$$e_1/e_2 = N_1/N_2 = i_2/i_1. \qquad (15.23)$$

Since the heat losses are negligible, the resistance of each winding must be negligible. Then in each winding the terminal p.d. equals the e.m.f., whatever the current, and the input power from the supply equals the output power from the secondary. Thus

$$P = V_1 I_1 = V_2 I_2, \qquad (15.24)$$

where V_1 and I_1 are the r.m.s. values of primary p.d. and current, and V_2 and I_2 are the corresponding secondary values.

Now, for a "load" of pure resistance R_2 across the secondary terminals, e_2 and v_2 are in the same phase as i_2, and since eqn. (15.23) applies at every instant, e_1, v_1, and i_1 are in the same phase also. Therefore, applying eqn. (15.23) to the r.m.s. levels,

$$V_1 = (N_1/N_2)V_2, \quad \text{and} \quad I_1 = (N_2/N_1)I_2 .$$

$$\therefore \; V_1/I_1 = (N_1/N_2)^2 V_2/I_2 = (N_1/N_2)^2 R_2 \qquad (15.25)$$

where R_2 is the secondary load resistance and $(N_1/N_2)^2 R_2$ is the apparent resistance "transferred" or "reflected" into the primary circuit.

Solution

(a) From eqn. (15.22) $N_1 = E_{m1}/(\omega \Phi_m)$, where the maximum primary e.m.f. $E_{m1} = \sqrt{(2)} \times 240$ V, $\omega = 2\pi f = 100\pi$ rad s^{-1}, $\Phi_m = 1.1 \times 10^{-3}$ Wb.

$$\therefore \text{ primary turns } N_1 = \sqrt{(2)} \times 240/(100 \times \pi \times 1.1 \times 10^{-3})$$
$$= 985. \qquad \text{(a)}$$

Then, using the transformer ratio, $N_2 = N_1 E_2/E_1$

$$= 985 \times 500/240 = 2050. \qquad \text{(a)}$$

(b) From eqn. (15.24), $I_1 = P/V_1 = 200/240 = 0.83$ A. (b)

and $I_2 = P/V_2 = 200/500 = 0.40$ A. (b)

(c) The actual load resistance $R_2 = V_2/I_2$

$$= 500/0.40 = 1250 \; \Omega. \qquad \text{(c)}$$

The resistance transferred to the primary

$$=V_1/I_1= 240/0.83 = 290\ \Omega. \qquad \text{(c)}$$

Alternatively, from eqn. (15.25),

$$V_1/I_1 = (985/2050)^2 \times 1250 = 290.$$

15.25. A transformer operating on a supply of 240 V, 50 Hz, has a core of cross-section 4.0×5.0 cm and primary and secondary windings of 600 and 1500 turns respectively. Calculate (a) the maximum flux density, and (b) the secondary e.m.f. [(a) 0.90 T; (b) 600 V.]

15.26. The maximum flux density of a transformer core of cross-section area 16 cm^2 is to be 1.0 T. Calculate the numbers of turns required for an input of 240 V, 50 Hz, and an output of 7.0 V on no load. [680, 20.]

15.27. A transformer has a primary winding of 1000 turns and a secondary of 50 turns. A resistor of 20 Ω is connected across the secondary terminals. Calculate the resistance transferred to the primary circuit. [8000 Ω.]

15.28. A 4 : 1 step down transformer with perfect coupling has core material requiring negligible magnetizing current, and primary and secondary windings of resistance 0.50 Ω and 0.22 Ω respectively. At an instant when the e.m.f. applied to the primary is 48 V, calculate the secondary terminal p.d. (a) on no load, and (b) with a secondary current of 8.0 A. [(a) 12 V; (b) 10 V.]

CHAPTER 16

A.C. CIRCUITS

Inductive Reactance

16.1. Worked Example

The p.d. across an inductor of negligible resistance is 10 V when passing a current of 20 mA at a frequency of 500 Hz. Calculate (a) the self-inductance, and (b) the current passed when an e.m.f. of 10 V at 50 kHz is applied.

Introduction

When an alternating current flows in an inductor, a back e.m.f. is induced, opposing the change of current and so causing a potential drop given by $L\,di/dt$, where L is the self-inductance and di/dt is the instantaneous rate of change of current i. Since the e.m.f. opposes the change of i, the potential drop v is a maximum V_m when i increases most rapidly, at the steepest parts of the graph of i against time t, as at $t = 0$ and T in Fig. 16.1. Similarly, v has a maximum negative value at $t = T/2$ and $3T/2$, etc. By considering the changing gradient of the sine wave of i in Fig. 16.1, the full variation of v may be seen to be the cosine wave, also shown.

Thus if $i = I_m \sin \omega t$,

$$v = V_m \cos \omega t = V_m \sin(\omega t + \pi/2). \tag{16.1}$$

Figure 16.2 shows the two phasors at the instant when $t = 0$ and i is increasing through zero, adopting the standard anticlock-

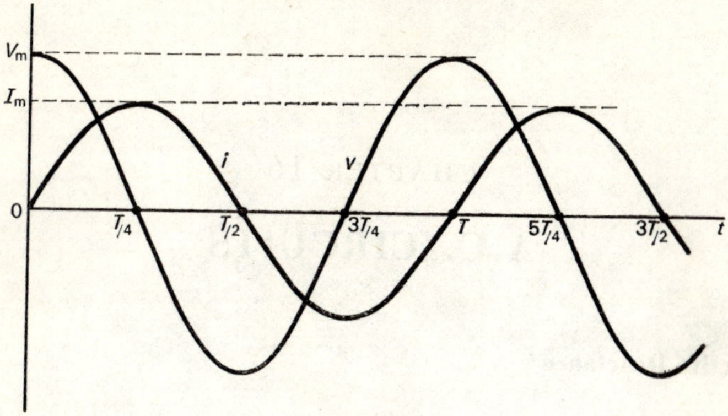

FIG. 16.1. Voltage and current in a pure inductor.

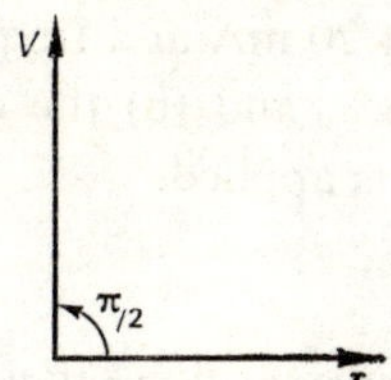

FIG. 16.2. Phasor diagram for a pure inductor.

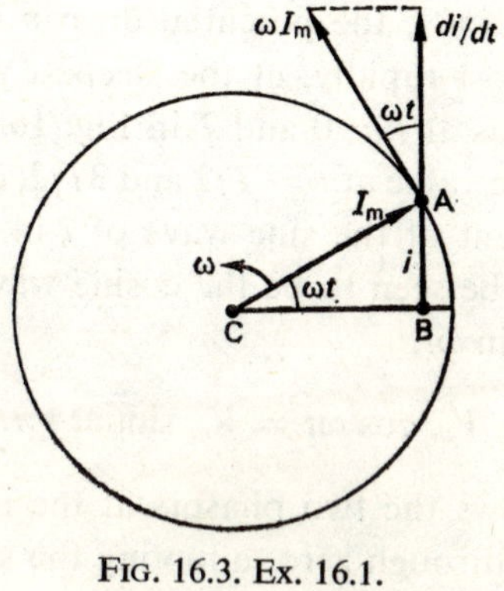

FIG. 16.3. Ex. 16.1.

wise direction of rotation (as in ex. 15.5). Usually, the symbols of r.m.s. values V and I are used instead of V_m and I_m on phasor diagrams, as Fig. 16.2, when comparing relative phases and magnitudes. However, in Fig. 16.3 the phasor CA represents I_m, not I_{rms}, so that the height of A above B represents $i = I_m \sin \omega t$, and the vertical velocity of A represents the rate of change di/dt.

Now the velocity of A around the circle = circumference ÷ period $= 2\pi CA/T = \omega CA$ representing ωI_m, and the vertical component of the velocity of $A = \omega CA \cos \omega t$.

$$\therefore\ di/dt = \omega I_m \cos \omega t.$$

Then $v = L\, di/dt = L(\omega I_m \cos \omega t) = V_m \cos \omega t$, as in eqn. (16.1). Therefore $V_m/I_m = \omega L$, and, equally,

$$V/I = \omega L = X_L, \qquad (16.2)$$

where V is the r.m.s. p.d. in volts across the inductor of pure inductance L in henrys, I is the r.m.s. current in amperes, $\omega = 2\pi f$ where f is the frequency in hertz (Hz), or cycles/second, and X_L is the inductive reactance in ohms.

Solution

(a) From eqn. (16.2), $X_L = V/I = 10/0.020 = 500\ \Omega$.

Also $\quad L = X_L/\omega = X_L/2\pi f = 500/(2\pi \times 500) = 0.16$ H. (a)

(b) At 50 kHz eqn. (16.2) gives $X_L = \omega L = 2\pi f L$

$$= 2\pi \times 50 \times 10^3 \times 0.16 = 5.0 \times 10^4\ \Omega.$$

Then $I = V/X_L = 10/(5 \times 10^4) = 2 \times 10^{-4}$ A

$= 0.20$ mA. (b)

Alternatively, note that eqn. (16.2) gives $I = V/(\omega L) = V/(2\pi f L)$. Therefore $I \propto 1/f$ for constant V and L. Thus when f changes by a factor of 100, from 500 Hz to 50 kHz, I changes by a factor of 1/100, from 20 mA to 0.20 mA.

16.2. Three inductors of inductance 10 H, 5.0 H, and 4.0 H respectively, are connected in series to a supply of e.m.f. 110 V at a frequency of 50 Hz. Calculate (a) the current, and (b) the p.d. across each inductor. [(a) 18 mA; (b) 58 V, 29 V 23 V.]

16.3. A current of 10 mA at a frequency of 2.5 kHz flows through an inductor of negligible resistance and of self-inductance 4.0 mH. Calculate (a) the maximum p.d., (b) r.m.s. p.d., and (c) the instantaneous ratio of p.d. to current at intervals of 1/30 ms throughout a positive half cycle of the p.d. [(a) 0.89 V; (b) 0.63 V; (c) 0, −36 Ω, −110 Ω, ∞, 110 Ω, 36 Ω, 0.]

Inductance and Resistance in Series

16.4. Worked Example

A sinusoidal e.m.f. of peak value 15.0 V and frequency 1.00 kHz is applied to a series circuit of resistance 750 Ω and self-inductance 0.100 H. Calculate (a) the impedance, (b) the r.m.s. value of the current, (c) the in-phase p.d., (d) the reactive p.d., (e) the Q-value, and (f) the instantaneous ratio of e.m.f. to current 750 μs after the e.m.f. rises positively through zero.

Introduction

When the resistance of an inductive circuit is considerable, the ratio of applied e.m.f. to current depends upon both the reactance X_L and resistance R. Although the resistance and inductance may be considered to be in series, as in Fig. 16.4, their resultant is not

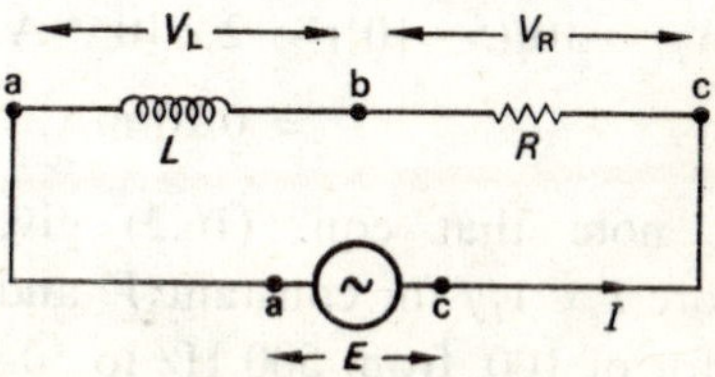

FIG. 16.4. Series a.c. R, L circuit.

a simple addition of X_L and R, even when both are due to the same coil. The resultant is found from the phasor diagram which is shown for three different instants in Figs. 16.5–16.7. In each case, I is drawn first since this is the same for each part of a series circuit: it is the reference phasor. Then the two p.d.s V_L and V_R are

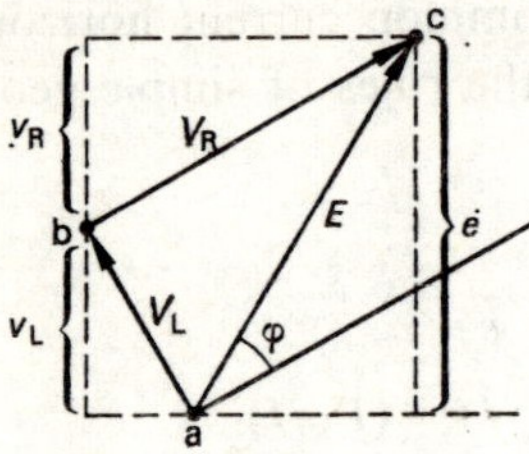

FIG. 16.5. Ex. 16.4.

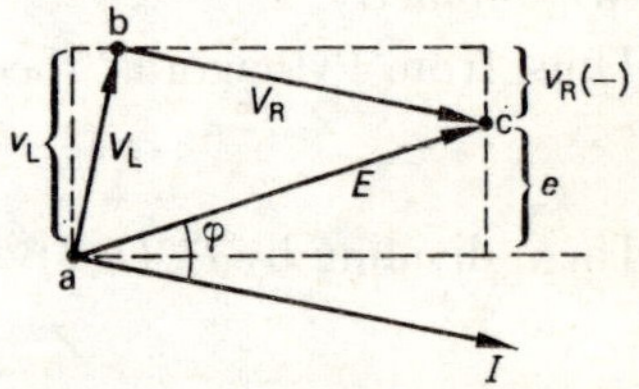

FIG. 16.6. Ex. 16.4.

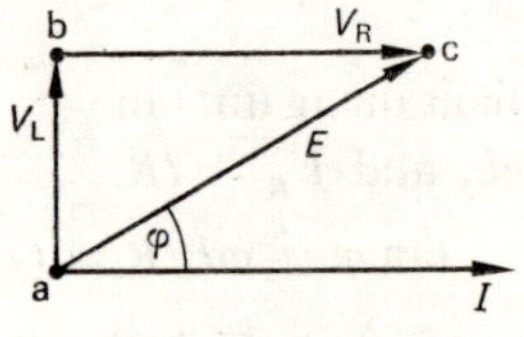

FIG. 16.7. Ex. 16.4.

shown as phasors in their correct phases relative to I, adding as in a triangle of vectors to give their resultant phasor E. V_R due to resistance is in phase with I, and V_L due to inductive reactance leads I "in quadrature", i.e. with 90° phase difference, as in ex. 16.1.

To distinguish between positive and negative p.d.s. in a series circuit, one direction is chosen to be positive for the current and indicated by an arrow head as in Fig. 16.4. Also the junctions of the circuit are lettered a, b, etc., taking the opposite direction to the current through the circuit from one supply terminal to the other. The phasors are then shown with the same letters a, b, etc., adding end to end in their correct phase relation to the current.

Then any point of the circuit is positive relative to another at the instant respresented, if the corresponding point on the phasor diagram is higher than the other; e.g. c is positive relative to b in Fig. 16.5, but negative in Fig. 16.6. Whatever the phase of I, the diagram would show $v_L+v_R = e$, where each of these instantaneous values may be positive or negative depending on phases. It is most convenient to show the common current horizontally, as in Fig. 16.7, and to calculate by the rules of simple geometry and trigonometry.

Thus, from Pythagoras' theorem,

$$E^2 = V_L^2+V_R^2. \qquad (16.3)$$

Then, dividing by I^2, $(E/I)^2 = (V_L/I)^2+(V_R/I)^2$.

$$\therefore\ Z^2 = X_L^2+R^2 \qquad (16.4)$$

and $Z = \sqrt{(X_L^2+R^2)}$, where $E/I = Z$, called the impedance of the circuit due to the resultant effect of resistance R in series with reactance X_L.

From the phasor diagram again, $\tan\varphi = V_L/V_R$. But from eqn. (16.2) $V_L = IX_L = I\omega L$, and $V_R = IR$.

$$\therefore\ \tan\varphi = \omega L/R = Q, \qquad (16.5)$$

where φ is the phase angle by which the e.m.f. leads the current, and Q is the "quality factor" by which an inductor is frequently judged when high reactance and low resistance are desirable.

Solution

(a) From eqn. (16.2), the reactance $X_L = \omega L = 2\pi fL$

$$= 2\pi\times10^3\times0.1 = 628\ \Omega.$$

Then, from eqn. (16.4), $Z^2 = X_L^2+R^2 = 628^2+750^2 = 9.58\times10^5$

$$Z = 978\ \Omega.$$

(b) Since $Z = E/I$, and the r.m.s. e.m.f. $E = 15/\sqrt{2} = 10.6$ V, the r.m.s. current $I = E/Z = 10.6/978 = 10.8\times10^{-3}$ A

$$= 10.8\ \text{mA}.$$

(c) The in-phase p.d. $V_R = IR = (10.8\times10^{-3})\times750 = 8.1$ V.

(d) The reactive p.d. $V_L = IX_L = (10.8\times10^{-3})\times628 = 6.8$ V.

As a partial check, $V_L^2+V_R^2 = 6.8^2+8.1^2 = 46+66 = 112 = 10.6^2 = E^2$.

(e) From eqn. (16.5), $Q = \omega L/R = 628/750 = 0.84$.

(f) Since $T = 1000$ μs when $f = 1$ kHz, 750 μs is three-quarters of a cycle. Then the instantaneous e.m.f. $e = -15$ V, as in Fig. 16.8. At the same instant $i = -I_m \cos\varphi$, where

$$I_m = \sqrt{(2)}I = 1.41\times10.8\times10^{-3}\text{ A} = 1.53\times10^{-2}\text{ A},$$

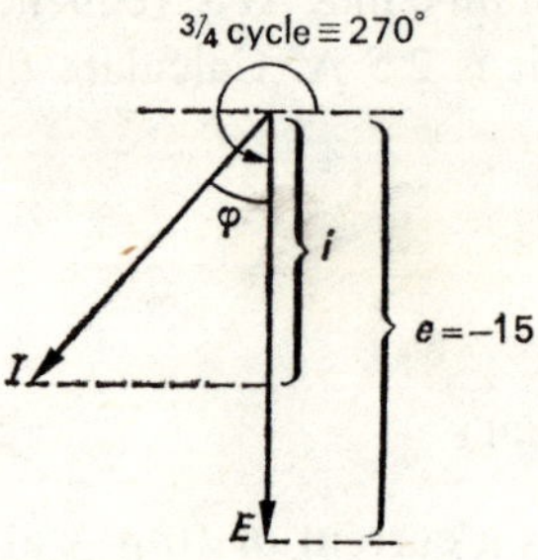

FIG. 16.8. Ex. 16.4(f).

and from Fig. 16.7,

$$\cos\varphi = V_R/E = IR/(IZ) = R/Z = 750/978.$$

$$\therefore \text{ at the stated instant } e/i = -15/(-1.53\times10^{-2}\times750/978) = 1280\ \Omega.$$

16.5. An e.m.f. of 10 V is applied to an inductor of series resistance 4.0 Ω and self-inductance 0.50 mH. Calculate the current, the in-phase p.d., and the reactive p.d. at frequencies of (a) zero, (b) 1.25 kHz, and (c) 5.0 kHz. [(a) 2.5 A, 10 V, 0; (b) 1.8 A, 7.1 V, 7.0 V; (c) 0.62 A, 2.5 V, 9.7 V.]

16.6. An inductor has a series resistance of 5.0Ω and an inductance of 2.0 mH. Calculate (a) the impedance at a frequency of

796 Hz, (b) the additional resistance required in series with the inductor to double the impedance at 796 Hz, and (c) the phase difference between the applied e.m.f. and current in each case (a) and (b). [(a) 11 Ω; (b) 15 Ω; (c) 63°, 26°.]

16.7. An inductive circuit draws an alternating current of 10.0 A at a p.d. of 250 V with the current lagging by a tenth of a cycle. Calculate the instantaneous current when the p.d. is (a) zero, and (b) peak. Also calculate the instantaneous p.d. when the current is (c) zero, and (d) peak. [(a) ±8.3 A; (b) ±11.4 A; (c) ±207 V; (d) ±285 V.]

16.8. An inductor is connected directly across a variable frequency supply of constant e.m.f. At a frequency of 50 Hz the current is 3.4 A; at 100 Hz it is 2.5 A. Calculate the current at 150 Hz. [1.9 A.]

Capacitive Reactance

16.9. Worked Example

A capacitor passes a current of 10 mA at a p.d. of 25 V and a frequency of 2.0 kHz. Calculate (a) the frequency at which the current would be 5.0 mA for a p.d. of 25 V, (b) the capacitance, and (c) the maximum charge held by the capacitor in each case.

Introduction

When an alternating e.m.f. is applied across a capacitor, an alternating current is drawn due to charge and discharge of the plates. The charge q does not pass through the dielectric, but equal currents i flow in the two leads, and the capacitor is said to pass the current i.

When the capacitor p.d. v is at a positive or negative peak it is momentarily constant; so also is q. Then i is zero, as when $t = T/4$ and $3T/4$ in Fig. 16.9. The charging current is greatest when the p.d. is rising most rapidly, as when $t = 0$ and T. Similarly, when

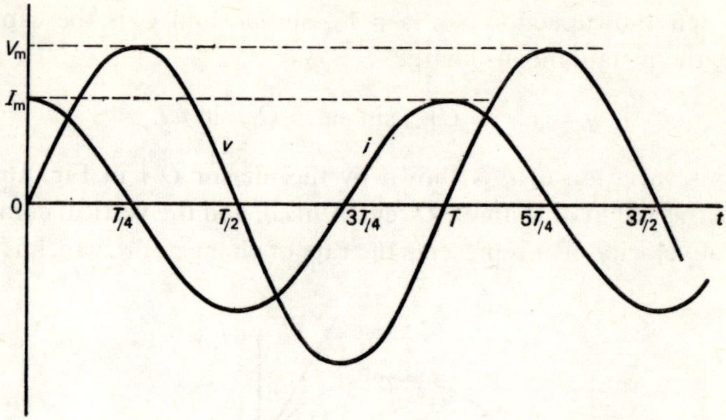

FIG. 16.9. Voltage and current in a pure capacitor.

the p.d. falls, the capacitor discharges with negative current. Thus i leads v, or v lags on i, by a quarter-cycle, and if $v = V_m$ sin ωt,

$$i = I_m \cos \omega t. \tag{16.6}$$

Figure 16.10 shows the phasors at two different instants, as at $t = 0$ in case (i), and as at $t = 3T/4$ in case (ii).

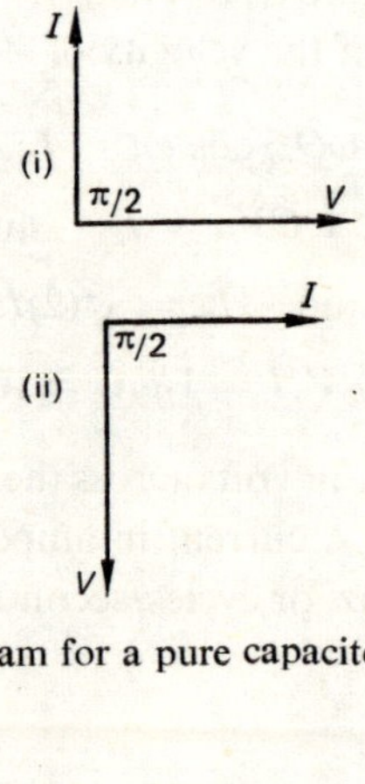

FIG. 16.10. Phasor diagram for a pure capacitor (i) $t = 0$, (ii) $t = 3T/4$.

When the capacitor p.d. $v = V_m \sin \omega t$, and C is the capacitance, the instantaneous charge

$$q = Cv = CV_m \sin \omega t = Q_m \sin \omega t.$$

This variation of q is shown by the phasor OA in Fig. 16.11, where the height of A above O represents q, and the vertical component of velocity of A represents the rate of change of q, which is the

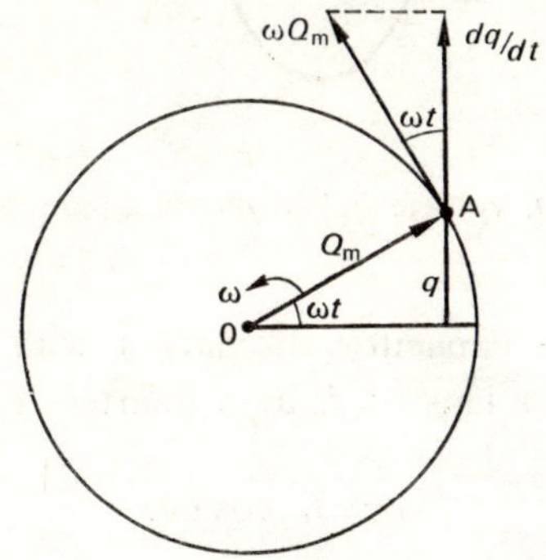

FIG. 16.11. Ex. 16.9.

current i, or dq/dt. Now the velocity of A around the circle = circumference ÷ period $= 2\pi OA/T = \omega OA$, representing ωQ_m, and the vertical component of the velocity of $A = \omega OA \cos \omega t$.

$$\therefore\ i = \omega Q_m \cos \omega t = I_m \cos \omega t. \tag{16.7}$$

But $\quad Q_m = CV_m. \quad \therefore\ \omega CV_m = I_m \quad$ and $\quad V_m/I_m = 1/\omega C.$

Now $\quad V_m = \sqrt{(2)}V, \quad$ and $\quad I_m = \sqrt{(2)}I.$

$$\therefore\ V/I = 1/\omega C = X_c, \tag{16.8}$$

where V is the r.m.s. p.d. in volts across the capacitor of capacitance C in farads, I is the r.m.s. current in amperes, $\omega = 2\pi f$, where f is the frequency in hertz, Hz, or cycles/second, and X_c is the capacitive reactance in ohms.

Solution

(a) From eqn. (16.8), $I \propto \omega \propto f$ for constant C and V.

Therefore the current changes from 10 mA to 5 mA when the frequency changes from 2 kHz to 1 kHz. (a)

(b) When $f = 2$ kHz, $V/I = 25/10^{-2} = 2500\ \Omega = X_C$. Then, from eqn. (16.8),

$$C = 1/(\omega X_C) = 1/(2\pi \times 2000 \times 2500) = 3.2 \times 10^{-8}\text{ F}. \qquad \text{(b)}$$

(c) Now $Q_m = CV_m$, where $V_m = \sqrt{(2)}V = 1.41 \times 25$ V.

$$\therefore\ Q_m = 3.2 \times 10^{-8} \times 1.41 \times 25 = 1.1 \times 10^{-6}\text{ C} = 1.1\ \mu\text{C} \qquad \text{(c)}$$

This applies to both frequencies since C and V are unchanged.

16.10. A capacitor of 0.10 μF is connected to a source of e.m.f. 15 V and frequency 300 Hz. Calculate the reactance and the current passed. [5.3 kΩ; 2.8 mA.]

16.11. Two capacitors connected in series have capacitances of 2.0 μF and 4.0 μF respectively. Calculate the current drawn and the p.d. across each capacitor when they are connected to a supply of e.m.f. 6.0 V, at a frequency of (a) 50 Hz, and (b) 2.5 kHz. [(a) 2.5 mA, 4.0 V, 2.0 V; (b) 130 mA, 4.0 V, 2.0 V.]

16.12. A current of 10 mA at a frequency of 2.5 kHz is supplied to a capacitor of 1.0 μF. Calculate (a) the maximum p.d., (b) the r.m.s. value of the p.d., and (c) the instantaneous ratios of p.d. to current at intervals of 1/30 ms throughout a positive half cycle of the p.d. [(a) 0.90 V; (b) 0.64 V; (c) 0, 37 Ω, 110 Ω, ∞, −110 Ω, −37 Ω, 0.]

Capacitor and Resistor in Series

16.13. Worked Example

An e.m.f. of 16 V at a frequency of 5.0 kHz acts in a circuit consisting of a capacitor of 0.10 μF in series with a resistor of 500 Ω. Calculate (a) the current, (b) the in-phase p.d., (c) the reactive p.d., and (d) the phase angle between the e.m.f. and current.

Introduction

The resultant impedance Z, due to the resistance R and the capacitive reactance X_c in series as in Fig. 16.12, is found by using a phasor diagram showing the addition of the p.d.s. Thus Fig. 16.13 serves the same purpose for C and R in series as Fig. 16.7 for R and L in series. Note that the capacitor p.d. V_c lags by a quarter of a

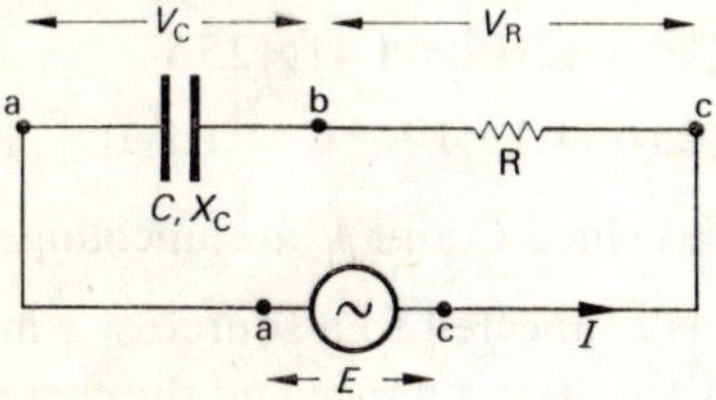

FIG. 16.12. Series a.c. R, C circuit.

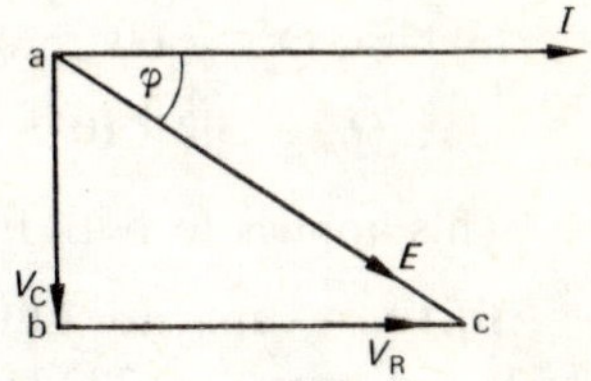

FIG. 16.13. Ex. 16.13.

cycle behind the current I, so that Fig. 16.13 is inverted in comparison with Fig. 16.7, but the geometrical structure is essentially the same.

Thus from Pythagoras' theorem,

$$E^2 = V_c^2 + V_R^2. \tag{16.9}$$

Then, dividing by I,

$$(E/I)^2 = (V_c/I)^2 + (V_R/I)^2.$$

$$\therefore\ Z^2 = X_c^2 + R^2 \tag{16.10}$$

and $Z = \sqrt{(X_C^2 + R^2)}$, where $E/I = Z$, called the impedance of the circuit due to the resultant effect of resistance R in series with reactance X_c.

Also from Fig. 16.13, $\tan\varphi = V_c/V_R$. (16.11)

Solution

(a) From eqn. (16.8) $X_c = 1/\omega C$, where $C = 0.1\times10^{-6}$ F and $\omega = 2\pi f = 2\pi 5000 = \pi\times10^4$ rad s^{-1}.

$$\therefore\ X_c = 1/(\pi\times10^4\times0.1\times10^{-6}) = 318\ \Omega.$$

Then, from eqn. (16.10), $Z^2 = 318^2+500^2 = 3.51\times10^5$.

$$\therefore Z = 593\ \Omega.$$

Then $I = E/Z = 16/593 = 2.7\times10^{-2}\text{ A} = 27\text{ mA}$. (a)

(b) The in-phase p.d. $V_R = I\times R$

$$= 2.7\times10^{-2}\times500 = 13\text{ V}. \quad \text{(b)}$$

(c) The reactive p.d. $V_c = I\times X_c$

$$= 2.7\times10^{-2}\times318 = 8.6\text{ V}. \quad \text{(c)}$$

As a partial check, note that $V_c^2+V_R^2 = 8.6^2+13.5^2 = 256$

$$= 16^2 = E^2,$$

as in eqn. (16.9).

(d) From eqn. (16.11), $\tan\varphi = 8.6/13.5 = 0.636$.

$$\therefore \varphi = 32^\circ. \quad \text{(d)}$$

16.14. A capacitor and a resistor are connected in series in a circuit carrying a current of 25 mA. The p.d. across the capacitor is 50 V and that across the resistor is 100 V. Calculate (a) the resistance, (b) the reactance, (c) the impedance, (d) the resultant p.d. across the two components, and (e) the phase angle between the resultant p.d. and the current. [(a) 4 kΩ; (b) 2 kΩ; (c) 4.5 kΩ; (d) 110 V; (e) 27°.]

16.15. A resistor of 50 Ω and a capacitor of 0.50 μF are connected in series with a supply of e.m.f. 100 mV at a frequency of 10 kHz. Calculate (a) the current drawn, (b) the p.d.s across the two components, and (c) the phase angle between the current and the e.m.f. [(a) 1.7 mA; (b) 85 mV, 54 mV; (c) 32°.]

16.16. A sinusoidal e.m.f. of peak value 10 V at a frequency of 1.0 kHz is applied to a series circuit of resistance 750 Ω and capacitance 0.50 μF. Calculate (a) the impedance, (b) the r.m.s. current, and (c) the instantaneous ratio of e.m.f. to current 250 μs after the current rises positively through zero. [(a) 810 Ω; (b) 8.7 mA; (c) 750 Ω.]

16.17. A p.d. v and a current i are specified as follows: $i = 24 \sin (1000\ \pi t)$ mA, and $v = 3.0 \sin (1000\ \pi t - \pi/3)$ V. Calculate i and v at an instant one-eighth of a cycle after v increases positively through zero. [23 mA; 2.1 V.]

16.18. A capacitor is in series with a resistor of 2.0 kΩ on a sinusoidal supply of e.m.f. 50 V. At a frequency of 20 Hz the capacitor p.d. is 31 V. Calculate the frequency at which the capacitor p.d. would be 36 V. [15 Hz.]

Power

16.19. Worked Example

Two chokes A and B are in series across a supply of e.m.f. 5.0 V at a frequency of 500 Hz. A has a resistance of 20 Ω and inductance of 4.0 mH. B has a resistance of 16 Ω and inductance of 2.0 mH. Calculate (a) the power dissipated by each choke, and (b) the resultant power factor of the complete circuit.

Introduction

In any circuit having one or more resistors in series or parallel the total power dissipated by the resistors

$$= I_1^2 R_1 + I_2^2 R_2 +, \text{ etc.,} \tag{16.12}$$

where each value of I is the r.m.s. current through that particular resistor, as in ex. 15.10. Alternatively, the various values of VI or V^2/R may be added, where each V is the r.m.s. p.d. across the particular resistor and I is the r.m.s. current through it.

In a complex circuit of various components, R, L, and/or C, the same method may be used, considering resistance only, wherever it occurs. Care must be taken that the correct values of I and V are considered with each R; e.g. in the case of a choke, the terminal p.d. V is the resultant of two components, only one of which is due to the resistance of the choke (inductor).

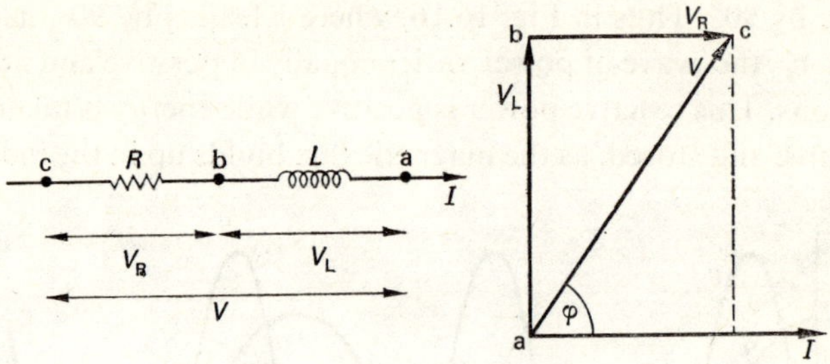

FIG. 16.14. Choke represented by R and L in series.

Thus in Fig. 16.14, the p.d. V_R, due to the resistance R, is $V \cos \varphi$.

$$\therefore \text{ power } P = V_R I = V \cos \varphi \; I, \tag{16.13}$$

where the factor ($\cos \varphi$) is called the power factor.

Equation (16.13) gives the mean value of the power, which fluctuates as in Fig. 16.15, where v_R and i are in phase and the instantaneous power p is obtained by multiplying v and i at numerous instants. In contrast to this it can be seen that a reactive component of p.d. accounts for no resultant power, whether leading or

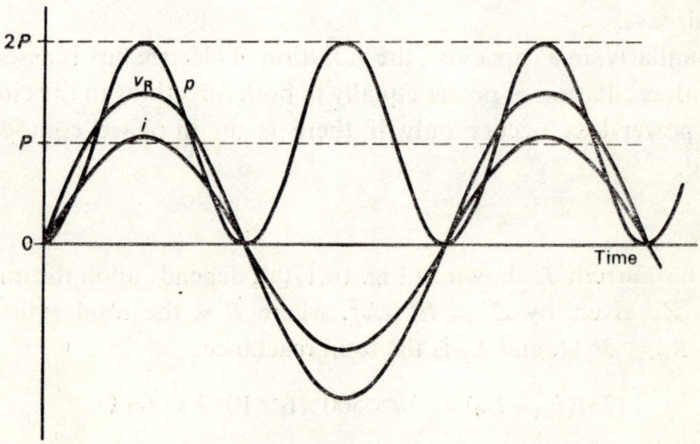

FIG. 16.15. Voltage, current, and power curves for a pure resistor.

lagging by 90°. Thus in Fig. 16.16, where v leads i by 90°, as in the case of v_L, the wave of power varies equally in positive and negative directions. This reactive power is positive while energy is taken from the source and stored, as the magnetic flux builds up in the inductor;

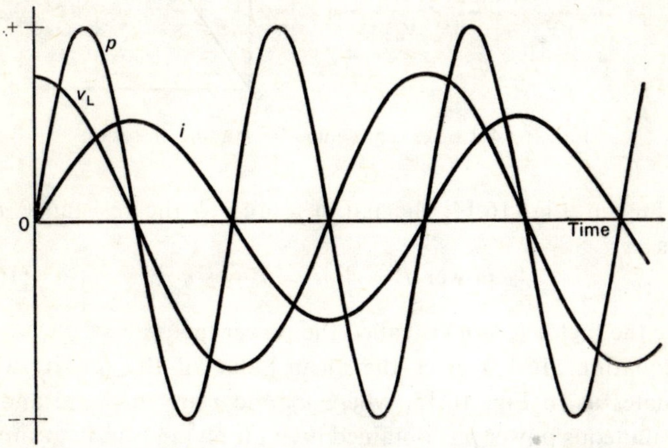

FIG. 16.16. Voltage, current, and power curves for a pure inductor.

it is negative while the energy is being returned to the source, as the flux decays.

Similarly, in a capacitor, the variation of electric flux is associated with oscillation of power equally in both directions in the circuit, and power loss occurs only if there is an in-phase component of p.d.

Solution

The current I, shown in Fig. 16.17(a), depends upon the impedance Z, given by $Z^2 = R^2+X_L^2$, where R is the total resistance $R_A+R_B = 36\ \Omega$, and X_L is the total reactance.

$$2\pi f(L_A+L_B) = 2\pi\times 500\times 6\times 10^{-3} = 6\pi\ \Omega.$$

Thus $Z^2 = 36^2+(6\pi^2) = 1655.$

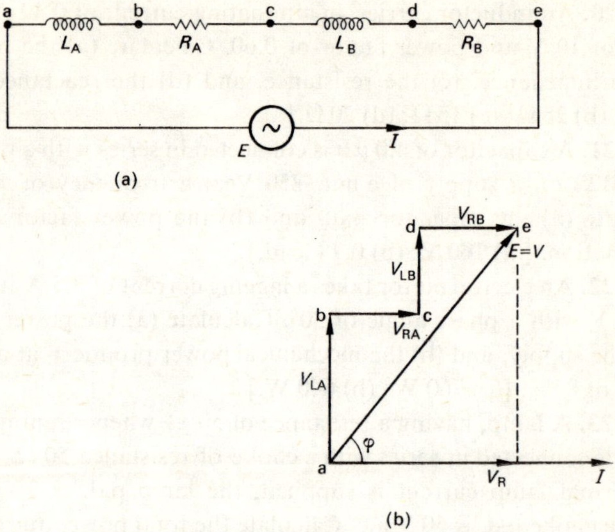

FIG. 16.17. (a) Two chokes in series (ex. 16.19). (b) Phasor diagram for ex. 16.19.

Then $I = E/Z = 5/\sqrt{1655} = 0.123$ A.

(a) For choke A,

$$P_A = I^2R_A = 0.123^2 \times 20 = 0.30 \text{ W} \qquad \text{(a)}$$

and for choke B,

$$P_B = I^2R_B = 0.123^2 \times 16 = 0.24 \text{ W}. \qquad \text{(a)}$$

(b) From (a) the total power $P_A + P_B = 0.54$ W.

But the overall p.d. $V = E = 5$ V, and the current $I = 0.123$ A. Therefore from eqn. (16.13) the power factor

$$\cos\varphi = P/(VI) = 0.54/(5 \times 0.123) = 0.88. \qquad \text{(b)}$$

Alternatively, without calculating power, $\cos\varphi$ may be calculated by the ratio V_R/V [see Fig. 16.17(b)].

Thus

$$\cos\varphi = I(R_A + R_B)/V = 0.123 \times 36/5 = 0.88. \qquad \text{(b)}$$

16.20. An inductor carries an alternating current of 0.40 A with a p.d. of 10 V and power factor of 0.60. Calculate (a) the power, (b) the impedance, (c) the resistance, and (d) the reactance. [(a) 2.4 W; (b) 25 Ω; (c) 15 Ω; (d) 20 Ω.]

16.21. A capacitor of 4.0 μF is connected in series with a resistor of 1000 Ω on a supply of e.m.f. 250 V at a frequency of 50 Hz. Calculate (a) the capacitor p.d., and (b) the power factor of the combination. [(a) 160 V; (b) 0.78 lead.]

16.22. An electric motor takes a lagging current of 3.5 A at a p.d. of 250 V with a phase angle of 30°. Calculate (a) the power taken from the supply, and (b) the mechanical power produced at an efficiency of 85%. [(a) 760 W; (b) 640 W.]

16.23. A lamp, having a resistance of 50 Ω when running normally, is connected in series with a choke of resistance 50 Ω. When the normal lamp current is supplied, the lamp p.d. is 25 V a.c. and the choke p.d. is 50 V a.c. Calculate the total power and power factor of the circuit. [25 W; 0.76 lag.]

Series Resonance

16.24. Worked Example

An inductor of resistance 200 Ω and inductance 10 mH is connected in series with a capacitor to give resonance at a frequency of 10 kHz. Calculate (a) the capacitance required, (b) the p.d. across the inductor when the supply e.m.f. is 5.0 V at the resonant frequency, and (c) the Q factor.

Introduction

In any series circuit the p.d.s across the components add to equal the applied e.m.f. at any instant, in accordance with Kirchhoff's second law (see ex. 3.27). The addition is shown by a phasor diagram indicating the r.m.s. values of the p.d.s and their phases relative to the common current. Thus in Fig. 16.18 the three p.d.s V_L, V_R, and V_C add in series to equal the applied e.m.f. E. The phase of each p.d. is related to the current as in Figs. 16.7 and 16.10, but the position of the phasor V_C has been moved without changing its angle relative to I. The p.d.s may be added in any order, but it is convenient to arrange the phasors with least overlap, as shown.

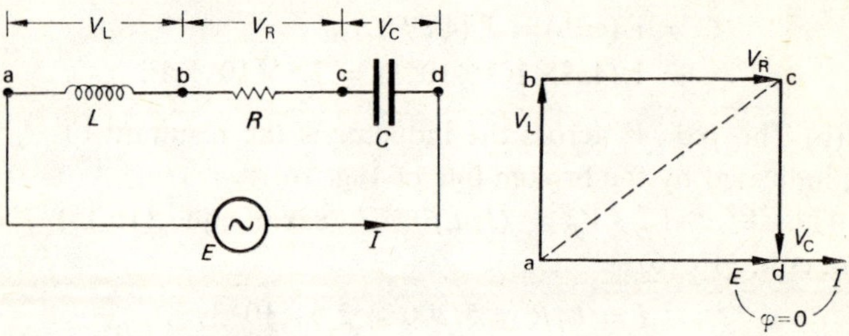

FIG. 16.18. Series R, L, C circuit and phasor diagram at resonance.

Series resonance of this circuit is the condition in which E and I are in phase, with $\varphi = 0$. Figure 16.18 shows that this occurs when

$$V_L = V_C \tag{16.14}$$

and

$$V_R = E. \tag{16.15}$$

But $V_R = IR$ and $E = IZ$.

$$\therefore\ Z = R \quad \text{and} \quad I = E/R \quad \text{at resonance.} \tag{16.16}$$

This gives the maximum current, since the resultant reactive p.d. is zero. Now, dividing (16.14) by I gives

$$X_L = X_C. \tag{16.17}$$

$$\therefore\ \omega L = 1/\omega C \quad \text{and} \quad \omega^2 = 1/CL. \tag{16.18}$$

Then the resonant frequency

$$f_r = \frac{\omega}{2\pi} = \frac{1}{2\pi}\sqrt{\left(\frac{1}{CL}\right)}. \tag{16.19}$$

The Q factor, as in ex. 16.4, is the ratio of reactive to resistive p.d., and is shown in Fig. 16.18 by the proportions of the rectangle when $\varphi = 0$.

Solution

(a) When $f_r = 10\times10^3$ Hz and $L = 10\times10^{-3}$ H, eqns. (16.18) and (16.19) give

$$C = 1/(\omega^2L) = 1/(4\pi^2 f_r^2 L)$$
$$= 1/(4\pi^2\times10^8\times10^{-2}) = 2.5\times10^{-8}\text{ F.} \qquad \text{(a)}$$

(b) The p.d. V across the inductor is the resultant of V_L and V_R, indicated by the broken line of Fig. 16.18.

Thus $V^2 = V_L^2 + V_R^2 = (I\omega L)^2 + E^2$, using eqn. (16.15). Now, from eqn. (16.16)

$$I = E/R = 5/200 = 2.5\times10^{-2}.$$
$$V^2 = (2.5\times10^{-2}\times2\pi\times10^4\times10^{-2})^2 + 5^2 = 247+25$$
$$\therefore V = \sqrt{(272)} = 16.5\text{ V.} \qquad \text{(b)}$$

(c) $Q = V_L/V_R = \omega L/R = 2\pi\times10^4\times10^{-2}/200 = \pi.$ (c)

16.25. A coil of self-inductance 25 mH is connected in series with a capacitor which is variable from 6.4 pF to 1000 pF. Calculate the range of resonant frequencies. [32 kHz to 0.40 MHz.]

16.26. A resistor of 50 Ω, an inductor of inductance 40 mH and negligible resistance, and a capacitor of 0.25 μF are connected in series to a supply of e.m.f. 15 V. Calculate (a) the frequency at which the current is a maximum, (b) the magnitude of this current, (c) the p.d. across the capacitor, and (d) the Q factor at this frequency. [(a) 1.6 kHz; (b) 0.3 A; (c) 120 V; (d) 8.]

16.27. An e.m.f. of 5.0 V at a frequency of 1.0 kHz is applied to a series circuit of three components: a resistor of 1 kΩ, a capacitor of 0.50 μF, and an inductor of 0.10 H with negligible internal resistance. Calculate (a) the additional capacitance required to obtain the maximum power, (b) the inductor p.d. when this is obtained, and (c) the Q-factor of the circuit. [(a) 0.52 μF, in series; (b) 3.1 V; (c) 0.63.]

General Series Circuit

16.28. Worked Example

An inductor of resistance 50.0 Ω and inductance 0.250 H is connected in series with a capacitor of capacitance 25.0 μF to a supply of e.m.f. 150 V at a frequency of 50.0 Hz. Calculate (a) the current, (b) the power, (c) the power factor, (d) the capacitor p.d., and (e) the inductor p.d.

Introduction

When a series circuit is not in resonance the current leads or lags on the supply e.m.f. to an extent found from the geometry of the phasor diagram. It is not known whether the current leads or lags in this particular question until a comparison has been made of X_L and X_C or V_L and V_C. The two possibilities are shown in Figs. 16.19 and 16.20, where the lettering and the relative phases are in accordance with the circuit diagram of Fig. 16.18.

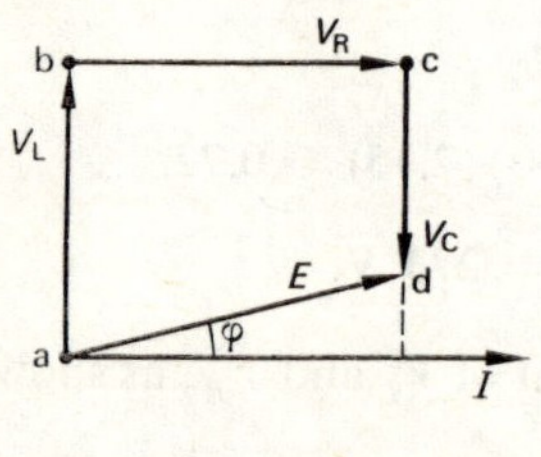

FIG. 16.19. Phasor diagram for series *R, L, C* circuit, current lagging.

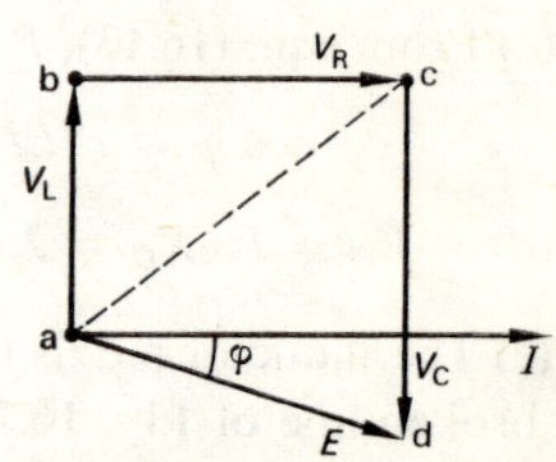

FIG. 16.20. Phasor diagram for series *R, L, C* circuit, current leading.

Now applying Pythagoras' theorem to the triangle containing the angle φ, whether leading or lagging,

$$E^2 = V_R^2+(V_L-V_C)^2. \tag{16.20}$$

Dividing eqn. (16.20) by I^2 gives

$$(E/I)^2 = (V_R/I)^2+(V_L/I-V_C/I)^2.$$

This simplifies to

$$Z^2 = R^2+(X_L-X_C)^2. \qquad (16.21)$$

Similarly, φ may be found from the geometry of the triangle.

Solution

(a) $X_L = 2\pi fL = 100\pi\times 0.25 = 78.6\ \Omega.$

$X_C = 1/(2\pi fC) = 1/(100\pi\times 25\times 10^{-6}) = 127.2\ \Omega.$

Now it is seen that since $X_C > X_L$, $V_C > V_L$, and the correct type of phasor diagram is Fig. 16.20.

From eqn. (16.21),

$$Z^2 = 50^2+(78.6-127.2)^2 = 2500+2370 = 4870$$

$$\therefore\ Z = \sqrt{(4870)} = 69.8\ \Omega.$$

Then $I = E/Z = 150/69.8 = 2.15\ \text{A}.$ (a)

(b) From eqn. (16.12), the power $P = I^2R = 2.15^2\times 50$

$= 232\ \text{W}.$ (b)

(c) From eqn. (16.13), $P = EI\cos\phi$.

$$\therefore\ \cos\varphi = P/EI = 232/(150\times 2.15) = 0.72. \qquad (c)$$

(d) $V_C = I\times X_C = 2.15\times 127.2 = 273\ \text{V}.$ (d)

(e) The inductor p.d. is the resultant of V_L and V_R, as shown by the broken line of Fig. 16.20.

Thus $V_{LR}^2 = V_L^2+V_R^2$, where $V_L = IX_L = 2.15\times 78.6 = 169$ V

and $V_C = IX_C = 2.15\times 127.2$

$= 107$ V.

$$\therefore\ V_{LR} = \sqrt{(169^2+107^2)} = \sqrt{40100} = 200\ \text{V}. \qquad (e)$$

16.29. A capacitor and an inductor, connected in series, draw a lagging current of 0.20 A from a supply of e.m.f. 10 V at a frequency of 1.5 kHz. The power drawn is 1.5 W. Calculate (a) the power factor, (b) the phase angle between the current and the e.m.f., (c) the resistance, and (d) the reactance of the circuit. If the inductance is 6 mH, calculate also the capacitance. [(a) 0.75; (b) 41°; (c) 38 Ω; (d) 33 Ω, 4.5 μF.]

16.30. An inductor of resistance 30 Ω and inductance 5.0 H is connected in series with a capacitor of 10 μF across a supply of e.m.f. 150 V at a frequency of 50 Hz. Calculate (a) the current, (b) the capacitor p.d., (c) the inductor p.d., (d) the in-phase and reactive components of the inductor p.d., and (e) the phase angle between the current and the e.m.f. [(a) 0.12 A; (b) 38 V; (c) 190 V; (d) 3.6 V, 188 V; (e) 89°.]

16.31. An inductor of resistance 20 Ω and inductance 2.0 mH is connected to a supply of e.m.f. 10 V at a frequency 5.0 kHz. Calculate the capacitance required in series (a) to halve the current, (b) to halve the power, (c) to halve the power factor, and (d) to halve the phase angle between the current and the e.m.f. [(a) 0.16 μF; (b) 0.21 μF; (c) 0.16 μF; (d) 2.2 μF.]

16.32. A simple filter circuit consists of an inductor and a capacitor in series across input terminals, with output terminals across the capacitor. The resistance and inductance of the inductor are 10 Ω and 100 mH respectively, and the circuit resonates at a frequency of 400 Hz. Calculate the ratio of output to input p.d. at a frequency of (a) 400 Hz and (b) 1200 Hz, assuming in each case that negligible output current is taken. [(a) 25; (b) 0.12.]

Resistor in Parallel with Pure Inductor or Capacitor

16.33. Worked Example

A capacitor of 0.10 μF is in parallel with a resistor of 1.0 kΩ on a supply of e.m.f. 5.0 V at a frequency of 796 Hz. Calculate (a) the supply current, and (b) the impedance.

Introduction

When two or more components are in parallel across the same source of e.m.f., the parallel currents may have different phase angles. To show the addition of these currents, the common e.m.f. is taken as the reference phasor, and usually shown horizontally as in Fig. 16.21. The branch currents are then shown in their correct phases relative to this e.m.f. E: in phase for I_R, leading in quadrature for I_C, lagging in quadrature for I_L and leading or lagging by some other definite angle for the current through a mixed impedance.

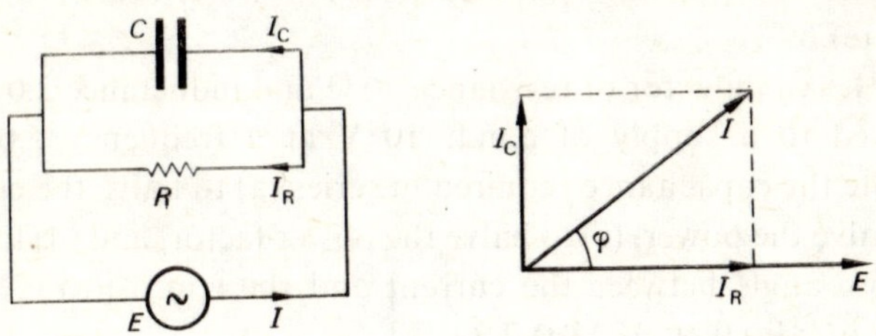

FIG. 16.21. Parallel a.c. *R*, *C* circuit and phasor diagram.

Thus Fig. 16.21 shows I_C leading I_R by 90°, and giving a resultant of I whose relative magnitude and phase angle may be calculated from the geometry of the diagram.

Applying Pythagoras' theorem,

$$I^2 = I_C^2 + I_R^2 . \tag{16.22}$$

Also,

$$\tan \varphi = I_C/I_R , \tag{16.23}$$

where $I = E/Z$, Z being the resultant impedance; $I_C = E/X_C$, where X_C is the reactance $1/\omega C$; and $I_R = E/R$.

Solution

(a) Using eqn. (16.22), $I^2 = (E/R)^2+(E/X_C)^2$
$= (E/R)^2+(E\times 2\pi fC)^2$
$= (5/1000)^2+(5\times 2\pi\times 796\times 10^{-7})^2$
$= 31.25\times 10^{-6}$.

$\therefore\ I = 5.6\times 10^{-3}\ \text{A} = 5.6\ \text{mA}.$ (a)

(b) $Z = E/I = 5/(5.6\times 10^{-3}) = 890\ \Omega.$ (b)

Alternatively, Z may be calculated from an equation for Z in terms of R, f, and C; i.e. $1/Z^2 = 1/R^2+4\pi^2f^2C^2$. However, it is usually more difficult to remember this equation than to adopt the method of (a) and (b), even when E is unknown. In the latter case, the symbol E would appear in the expression for the current, but would cancel out of the value for Z.

16.34. A capacitor of 0.10 μF is connected in parallel with a resistor of 1.0 kΩ to a supply of e.m.f. 15 V at a frequency of 1.5 kHz. Calculate the supply current and its phase angle. [21 mA; 43° leading.]

16.35. Calculate (a) the capacitance of the capacitor required in parallel with a resistor of 500 Ω to draw a total current of 55 mA from a supply of e.m.f. 25 V at 50 Hz. Calculate also (b) the phase angle of the resultant current, (c) the reactance, and (d) the current drawn by the capacitor. [(a) 2.9 μF; (b) 25°; (c) 1.1 kΩ; (d) 23 mA.]

16.36. A resistor of 20 Ω is connected in parallel with an inductor of 10 mH and negligible resistance to a supply of e.m.f. 2.0 V at 500 Hz. Calculate the current (a) in the resistor, (b) in the inductor, and (c) in the supply leads. Calculate also the phase angle of the current in the supply leads relative to the e.m.f. [(a) 0.10 A; (b) 0.064 A; (c) 0.12 A, 32° lagging.]

16.37. A certain imperfect capacitor at a frequency of 796 Hz is equivalent to a pure capacitor of 2.0 μF in series with a resistor of 50 Ω. Calculate the capacitance and resistance in parallel which would give the same impedance and power factor at the same frequency. [1.6 μF; 250 Ω.]

Capacitor in Parallel with Inductor

16.38. Worked Example

A supply of sinusoidal e.m.f. 10 V is connected across a parallel circuit consisting of a capacitor of 1.0 μF in parallel with an inductor of series resistance 50 Ω and self-inductance 12 mH. Calculate (a) the current in the inductor when the frequency is 1.5 kHz, (b) the current from the supply at 1.5 kHz, and (c) the current from the supply at the resonant frequency.

Introduction

The branch currents I_{LR} and I_C, shown in Fig. 16.22, are to be added by means of a phasor diagram with E as the reference phasor. However, the current I_{LR} lags E by an angle φ_{LR} which is shown

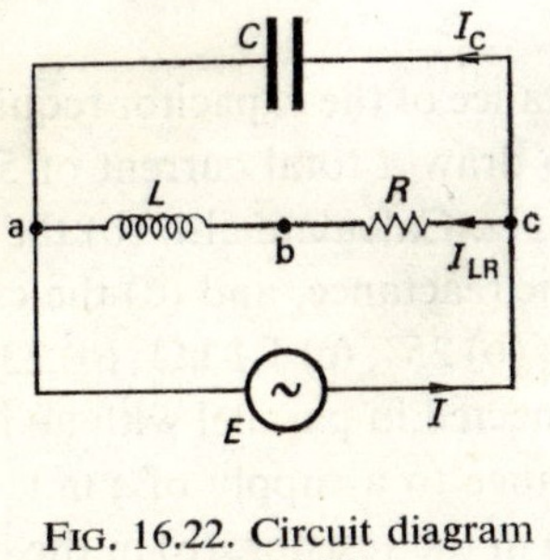

Fig. 16.22. Circuit diagram (ex. 16.38).

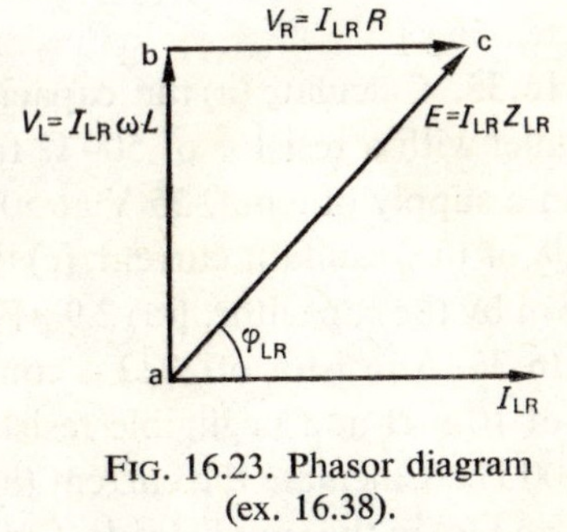

Fig. 16.23. Phasor diagram (ex. 16.38).

first in Fig. 16.23, obtained by considering the inductive branch alone as a series circuit. φ_{LR} is then used in constructing the main diagram which is based on E and has one of three forms, depending on φ_{LR} and the relative magnitudes of the branch currents. Thus Fig. 16.24 applies at the higher frequencies, for which $I_C > I_{LR} \sin \varphi_{LR}$; Fig. 16.25 at the lower frequencies for which $I_C < I_{LR}$

$\sin \varphi_{LR}$; Fig. 16.26 at the critical frequency of resonance, for which $I_C = I_{LR} \sin \varphi_{LR}$, and therefore I is in phase with E.

In general,

$$I \cos \varphi = I_{LR} \cos \varphi_{LR} \tag{16.24}$$

$$I \sin \varphi = I_C \sim I_{LR} \sin \varphi_{LR}. \tag{16.25}$$

$$\therefore\ I^2 = (I_{LR} \cos \varphi_{LR})^2 + (I_C \sim I_{LR} \sin \varphi_{LR})^2. \tag{16.26}$$

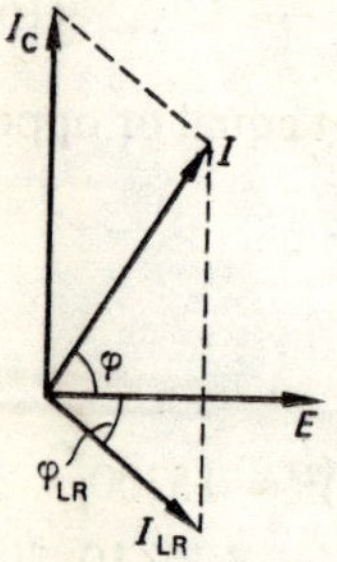

FIG. 16.24. High frequency, current leads (ex. 16.38).

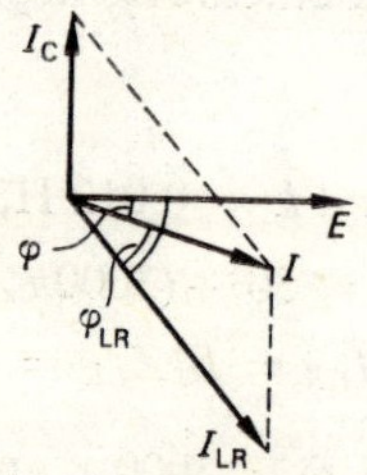

FIG. 16.25. Low frequency, current lags (ex. 16.38).

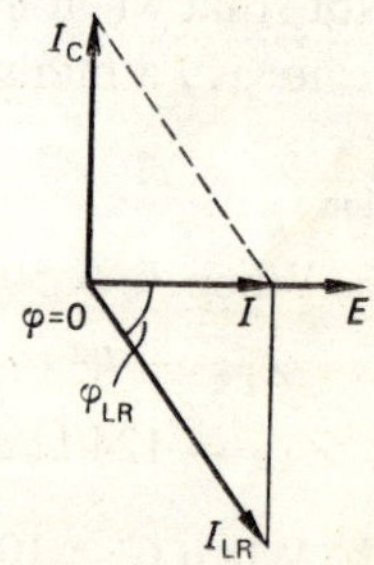

FIG. 16.26. Resonant frequency, current in phase with voltage (ex. 16.38).

At resonance, when $\varphi = 0$, eqn. (16.24) gives

$$I = I_{LR} \cos \varphi_{LR} \tag{16.27}$$

and eqn. (16.25) gives

$$I_C = I_{LR} \sin \varphi_{LR}. \tag{16.28}$$

But from Fig. 16.23, $\cos \varphi_{LR} = R/Z_{LR}$, $\sin \varphi_{LR} = \omega L/Z_{LR}$, and $Z_{LR}^2 = R^2 + \omega^2 L^2$. Also $I = E/Z_{LR}$, and $I_C = E\omega C$. Substituting these expressions in eqn. (16.28) gives an expression for ω_r at resonance. Thus eqn. (16.28) becomes $E\omega_r C = (E/Z_{LR}) \times (\omega_r L/Z_{LR})$.

$$\therefore\ L/C = Z_{LR}^2 = R^2 + \omega_r^2 L^2. \tag{16.29}$$

$$\therefore\ \text{at resonance } \omega_r^2 = 1/CL - R^2/L^2. \tag{16.30}$$

Substituting in eqn. (16.27) gives an expression for the resultant impedance at resonance, called the "dynamic impedance" Z_r. Thus eqn. (16.27) becomes

$$E/Z_r = (E/Z_{LR})\times(R/Z_{LR}).$$

Therefore $Z_r = Z_{LR}^2/R$. Then, with eqn. (16.29),

$$Z_r = L/(CR). \tag{16.31}$$

Note that when $\varphi = 0$ at resonance, I_{LR} is not equal or opposite to I_C, nor is I a minimum, unless R is negligible.

Solution

(a) When $R = 50\,\Omega$ and $L = 0.012$ H,

$$Z_{LR}^2 = R^2+\omega^2L^2 = 50^2+(3000\pi\times0.012)^2 = 15300.$$

$$\therefore\ Z_{LR} = 124\ \Omega \text{ and } I_{LR} = E/Z_{LR} = 10/124 = 8.1\times10^{-2}\text{ A}.$$

(b) When $C = 10^{-6}$ F, $\omega = 3000\,\pi$, and $E = 10$ V.

$$\therefore\ I_C = E\omega C = 10\times3000\pi\times10^{-6} = 9.4\times10^{-2}\text{ A}.$$

Now I is given by eqn. (16.26), in which

$$\sin\varphi_{LR} = \omega L/Z_{LR} = (3000\pi\times0.012)/124 = 0.913,$$

and $$\cos\varphi_{LR} = R/Z_{LR} = 50/124 = 0.404.$$

Then eqn. (16.26) gives

$$I^2 = (8.1\times10^{-2}\times0.404)^2+[(9.4\times10^{-2})-(8.1\times10^{-2}\times0.913)]^2$$
$$= (10.7\times10^{-4})+(4\times10^{-4}).$$
$$\therefore\ I = \sqrt{(14.7\times10^{-4})} = 3.8\times10^{-2}\text{ A}.$$

(c) At the resonant frequency, eqn. (16.31) gives

$$Z_r = L/(CR) = 0.012/(10^{-6}\times50) = 240\ \Omega.$$

Then at resonance $I_r = E/Z_r = 10/240 = 4.2\times10^{-2}$ A. Note that the current in (c) is greater than in (b) although (c) is at resonance.

16.39. Calculate (a) the resonant frequency of the circuit of ex. 16.38, and (b) the resonant frequency of a circuit of the same inductance and capacitance but of negligible resistance. [(a) 1.3 kHz; (b) 1.5 kHz.]

16.40. A parallel-resonant circuit consists of a capacitor in parallel with an inductor of inductance 5.0 mH and resistance 200 Ω. Calculate the capacitance required for resonance at a frequency of 5.0 kHz. If the circuit is connected to a supply of e.m.f. 10 V at this frequency, calculate the supply current and power dissipated. [0.077 μF; 31 mA; 0.31 W.]

16.41. An inductor of resistance 100 Ω and inductance 5.0 H is connected in parallel with a capacitor of 1.5 μF to a supply of e.m.f. 25 V at a frequency of 50 Hz. Calculate the current in (a) the inductor, (b) the capacitor, and (c) the supply leads, and calculate (d) the phase angle of the supply current, and (e) the power drawn. [(a) 16 mA; (b) 12 mA; (c) 4.2 mA; (d) 76°; (e) 25 mW.]

Simple Bridge Circuits

16.42. Worked Example

A bridge for the measurement of an imperfect capacitor is balanced when the arm opposite to this capacitor consists of a resistor of 400 Ω, and the other arms consist of (1) a resistor of 150 Ω and (2) a series combination of 12 Ω resistance and 360 pF capacitance. Calculate the equivalent series resistance and capacitance of the capacitor under test, and its power factor at a frequency of 1.0 kHz.

Introduction

A bridge circuit is balanced when its two potential dividers give equal output p.d.s. Thus in Fig. 16.27, Z_1 and Z_4 form one potential divider, Z_2 and Z_3 form the other, and the bridge is balanced when the potentials of N and L are equal. If the supply is of audio frequency, balance may be detected by phones or other a.c. detector D, bridging across from N to L. Balance occurs when the potentials of N and L are equal at all instants, i.e. when both the magnitudes and the phases of these potentials are equal.

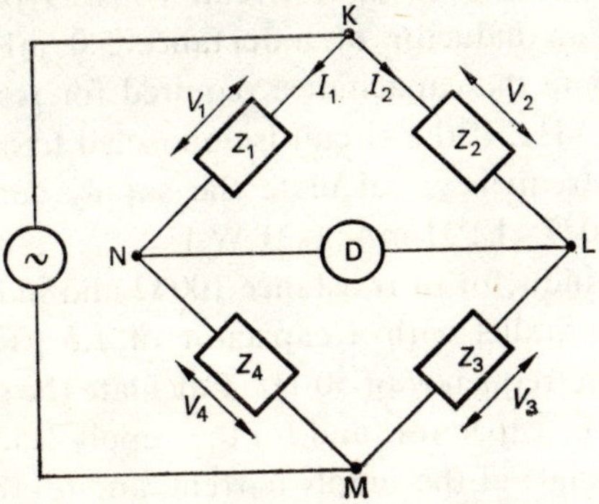

FIG. 16.27. Four-arm a.c. bridge circuit.

Thus if Z_1 and Z_2 are impedances giving the same phase angle, e.g. both resistors, or both capacitors, then the currents I_1 and I_2 must be in the same phase to make the p.d.s V_1 and V_2 also in the same phase at balance, when

$$V_1 = V_2. \tag{16.32}$$

Similarly,

$$V_3 = V_4 \tag{16.33}$$

in magnetude and phase when Z_3 and Z_4 are impedances of the same phase angle.

Now no current passes through the detector at balance.

$$\therefore \; V_4 = I_1Z_4 \quad \text{and} \quad V_3 = I_2Z_3.$$

Substituting in eqn. (10.33):

$$I_1Z_4 = I_2Z_3. \tag{16.34}$$

Dividing eqn. (16.32) by eqn. (16.34):

$$V_1/(I_1Z_4) = V_2/(I_2Z_3)$$

$$\therefore \; \frac{Z_1}{Z_4} = \frac{Z_2}{Z_3}, \tag{16.35}$$

where, if one pair of adjacent impedances give equal phase angles, so must the other pair. Note that the positions of the impedances in the circuit are similar to those in eqn. (16.35). The supply and the detector may be interchanged without affecting the relationship of the impedances at balance if stray effects are negligible, but calculation may be easier when considering one arrangement rather than the other.

Solution

If Z_1 is the imperfect capacitor under measurement, Z_3 is the 400 Ω resistor, and Z_4 may be the 150 Ω resistor. Then Z_2 is the impedance of the 12 Ω resistance in series with the 360 pF capacitance.

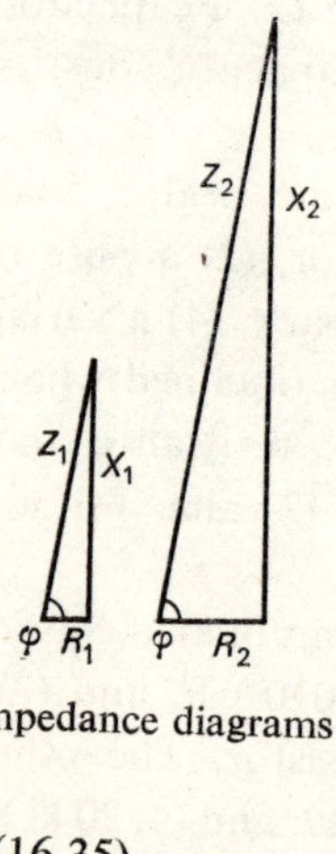

FIG. 16.28. Impedance diagrams for ex. 16.42.

Therefore, from eqn. (16.35),

$$Z_1 = Z_2Z_4/Z_3 = Z_2 \times 150/400.$$

Now Z_1 and Z_2 must give the same phase angle φ, as in Fig. 16.28.

$$\therefore \; R_1 = R_2 \times 150/400$$
$$= 12 \times 150/400 = 4.5 \; \Omega$$

and

$$X_1 = X_2 \times 150/400.$$

$$\therefore\ 1/\omega C_1 = (1/\omega C_2)\times 150/400.$$

$$\therefore\ C_1 = C_2\times 400/150$$

$$= 360\times 400/150 \text{ pF} = 960 \text{ pF}.$$

Then the power factor $\cos\varphi = R_1/Z_1$, where

$$Z_1 = \sqrt{(R_1^2+X_1^2)} \quad \text{and} \quad X_1 = 1/\omega C_1 = 10^{12}/(2\pi\times 10^7\times 960)$$

$$= 1.65\times 10^5\ \Omega.$$

Thus R_1^2 is negligible in comparison with X_1^2, so that

$$\cos\varphi = R_1/X_1 = 4.5/(1.65\times 10^5) = 2.7\times 10^{-5}.$$

16.43. Calculate the capacitance of a capacitor which balances in an a.c. bridge circuit when the neighbouring arms consist of (1) a resistor of 750 Ω, and (2) a capacitor of 12 μF, and the arm opposite to the unknown capacitor consists of a resistor of 1750 Ω. [28 μF.]

16.44. Four arms of an a.c. bridge, taken in cyclic order, consist of (1) an unknown inductor, (2) a pure resistor of 1000 Ω, (3) a variable non-inductive resistor, (4) a variable inductor of constant resistance 3.6 Ω. Balance is obtained when the variables are 450 Ω and 27 mH. Calculate the resistance and inductance of the unknown inductor, and its Q-value for a frequency of 1.0 kHz. [8 Ω; 60 mH; 47.]

16.45. Two adjacent arms of an a.c. bridge circuit consist of (1) a capacitor nominally of 0.010 μF, and (2) a variable capacitor in series with a variable resistor. The other arms, taken in cyclic order, are resistors (3) 10 kΩ, and (4) 20 kΩ. Calculate the equivalent parallel capacitance and resistance of the capacitor under test, when balance at a frequency of 1592 Hz occurs with $R = 52\ \Omega$ and $C = 0.022$ μF. [1.1×10^{-8} F; 0.80 MΩ.]

INDEX